A Field Guide to California Lichens

A FIELD GUIDE TO

California Lichens

Stephen Sharnoff

Foreword by Peter H. Raven

Yale UNIVERSITY PRESS

New Haven and London

Published with assistance from the foundation established in memory of Amasa Stone Mather of the Class of 1907, Yale College.

Yale University Press books may be purchased in quantity for educational, business, or promotional use. For information, please e-mail sales.press@yale.edu (US office) or sales@yaleup.co.uk (UK office).

Epigraph: "For the *Lobaria, Usnea,* Witches Hair, Map Lichen, Beard Lichen, Ground Lichen, Shield Lichen," from *Come Thief: Poems* by Jane Hirshfield, copyright © 2011 by Jane Hirshfield. Used by permission of Alfred A. Knopf, a division of Random House, Inc.

Designed by Nancy Ovedovitz and set in Chapparal Pro and Meta Plus types by BW&A Books, Inc. Printed in China.

Library of Congress Cataloging-in-Publication Data
Sharnoff, Stephen, 1944–
A field guide to California lichens / Stephen Sharnoff ;
foreword by Peter H. Raven.
pages cm
Includes bibliographical references and index.
ISBN 978-0-300-19500-2 (paperbound : alk. paper)
1. Lichens—California—Identification. 2. Lichens—
California—Pictorial works. I. Title.
QK587.5.C2S53 2014
579.7—dc23   2013041274

A catalogue record for this book is available from the British Library.

This paper meets the requirements of ANSI/NISO Z39.48–1992 (Permanence of Paper).

10 9 8 7 6 5 4 3 2 1

To the memory of Sylvia Sharnoff,

who began the adventure

For the *Lobaria, Usnea,* Witches Hair, Map Lichen, Beard Lichen,
Ground Lichen, Shield Lichen

Back then, what did I know?
The names of subway lines, busses.
How long it took to walk twenty blocks.

Uptown and downtown.
Not north, not south, not you.

When I saw you, later, seaweed reefed in the air,
you were gray-green, incomprehensible, old.
What you clung to, hung from: old.
Trees looking half-dead, stones.

Marriage of fungi and algae,
chemists of air,
changers of nitrogen-unusable into nitrogen-usable.

Like those nameless ones
who kept painting, shaping, engraving
unseen, unread, unremembered.
Not caring if they were no good, if they were past it.

Rock wools, water fans, earth scale, mouse ears, dust,
ash-of-the-woods.
Transformers unvalued, uncounted.
Cell by cell, word by word, making a world they could live in.

—Jane Hirshfield

Contents

- - - - - - - - -

Foreword

\- - - - - - - -

Lichens are wonderfully curious organisms, fungi that, by virtue of their usually tough, drought-resistant tissues and the photosynthetic organisms that grow within them, are enabled to colonize a wide variety of habitats where they could not otherwise survive. Some lichens are conspicuous elements in the habitats where they grow and thus familiar to everyone—as gray-green or yellowish green hanging masses on tree branches; orange, yellow, red splotches on rocks; or in one of the other forms illustrated in this book—but many lichens are smaller or have inconspicuous colors. Such lichens may easily go unnoticed by a general observer. All but a very few kinds of lichens are ascomycetes, a major group of fungi that includes morels, cup fungi, yeasts, and the important laboratory organism *Neurospora*. Lichens reproduce and are often dispersed by spores in the same ways as other ascomycetes, sexually or asexually, or sometimes mechanically, when pieces of tissue simply break off the parent organism. In the last case, the included photosynthetic cells may be dispersed as a part of the mass, along with the fungal units being dispersed; if not, the fungi and algae get joined when they are both present on the substrate where the lichen will grow. Lichen fungi are diverse, not simply fungi that have been transformed into lichens by taking up photosynthetic organisms, but distinct families, genera, and species of fungi that always live in this kind of symbiotic association. Although there are many tens of thousands of species of fungi that live as lichens worldwide, only a few hundred kinds of photosynthetic bacteria or algae occur in association with them. The same ones thus occur with many kinds of fungi to form many kinds of lichens.

Those lichens that harbor photosynthetic bacteria (cyanobacteria, or blue-green "algae") are important in converting nitrogen from the atmosphere into forms where it can be used by the fungi and, washed out by rain, effectively fertilize whole forests, like the dense coniferous forests in coastal regions. Crusting lichens are important in the first stages of slowly reducing rocks to soil, gradually growing over the rocks and etching and ultimately breaking down their surfaces, thus forming a roothold for mosses, liverworts, and other land plants that could not otherwise grow there. Other kinds of lichens grow on the branches of trees or other plants, on walls, on dirt—virtually in every natural habitat. On building stones, in gardens, or in cemeteries, they often provide a touch of color or the appearance of antiquity that can add a picturesque accent to the view. Within the tissues of lichens are produced many kinds of unique chemicals that are found nowhere else. These lichen chemicals are being studied actively in laboratories all over the world because of their intrinsic interest and usefulness in classifying the organisms that produce them. Certain kinds of lichens are consumed as food by people in some parts of the world, while others are important sources of dyes used in weaving, such as Harris tweed. At high latitudes or high elevations, lichens, some of which are called reindeer moss, may form major sources of food for grazing mammals. Lichens can be particularly obvious elements in the low vegetation of such places.

California is rich in lichen species, more than fifteen hundred of them recorded so far and doubtless many more remaining to be found. As Stephen Sharnoff emphasizes in the Introduction, anyone who looks carefully at lichens in the state will soon find species not described or illustrated in this book, some of which will actually prove to be new to the region or even new to science. Lichens in different ways play important roles in the ecology of terrestrial communities all over California, occurring in every conceivable habitat and eco-region. They hold a special interest for those who enjoy the out-of-doors by virtue of their unique biological nature and varied forms and colors; there is a great deal to be learned about them through further study. In general, we have not had field guides such as this one available that allow interested students to find and identify our lichens; this superbly illustrated book thus fills a definite void for naturalists throughout California. Stephen Sharnoff has done an amazing job of photographing the very wide variety of species illustrated in this manual, his efforts making it as simple as possible, with the aid of a hand lens, to identify the various lichens that we encounter. His earlier coauthored book, *Lichens of North America,* illustrates a wide range of lichens from throughout the United States and Canada with beautiful and instructive images taken by him and his late wife, Sylvia.

That book also includes a number of general chapters that contain a wealth of additional information about these fascinating organisms: it is highly recommended for anyone interested in learning more about lichens. In general, the distributions of particular kinds of lichens in California are similar to those of other organisms in the state, with each distinctive region having its own special complement of genera and species. Like many plants that grow in arid areas, especially ones near the coast, many kinds of lichens collect moisture from fog, much of which drips to the ground, thus greatly increasing the effective precipitation in regions that are apparently very dry, like the coastal sage associations of Southern California and the foggy deserts of northern Baja California.

In not generally growing in urban or other polluted areas, lichens are bellwethers for what is happening to the overall environment. Air and water pollution of various kinds limits their growth just as it harms human health and gradually decreases biological diversity. When I was a child in California in the 1930s, the population of the state amounted to about six million people; now there are nearly forty million there. For the world as a whole, there are now three people living for each one who was there when I was born! Now and for the next several decades worldwide, people will continue to be added at the rate of about two hundred thousand net each day. The population of California is projected to keep rising rapidly as well, and expectations for ever-higher levels of consumption seeming to be almost endless. The integrity of natural communities continues to disintegrate as more land is cleared, more pollution is added, invasive species increase in number and kind, and climate change proceeds more rapidly than we can sense within the limits of our relatively short lives.

For these reasons, it has been estimated that of the more than two thousand species of land plants restricted to California, half or more will be extinct or severely restricted in range by the end of the century. Lichens and other organisms will be no different, and understanding them will provide a key to the health of natural ecosystems as the years go by and perhaps provide clues to the most appropriate ways to preserve the habitats where they occur. Lichens have not, in general, been cultivated successfully, and their communities can be expected to be disturbed deeply by environmental changes in the future. For this reason, they should be studied carefully and conserved to the extent possible in the years to come.

In their diversity of color, form, and the habitats where they grow, lichens are a fascinating and important part of the web of life that will amply repay study at any level. By observing the world around us, we become better and more responsible keepers of the sustainable systems on which we depend. For observant children, especially, the world can become a wonderful place that opens ever more before them and forms the basis for a lifelong love of nature, filled with interest and variety. All Californians should be grateful for the appearance of this fine guide, a welcome introduction to the study of abundant but often unnoticed organisms that play such critical roles in the balance of nature.

Peter H. Raven

Preface

- - - - - - - -

Sometime in the 1970s, Sylvia Sharnoff and I visited a local California museum. We stood in front of a diorama of the alpine Sierra Nevada, a carefully detailed model of rugged mountains with a foreground of large granite boulders, enlivened by a furry stuffed marmot perched on a rock and a large bird poised in flight overhead. It was not some generic reconstruction but a view of an actual place to which we had often backpacked, Little Lakes Valley. Yet something was deeply wrong—why did the scene look dull and unnatural? It took some time to figure it out: they forgot the lichens! Without the colors and textures that lichens add, the rocks could have been on the moon rather than part of our living planet. Lichens are a visible and important component of the landscape, yet they are often taken for granted or assumed to be intrinsic to what they grow on, not perceived as lifeforms in their own right. This book provides an introduction to these remarkable and diverse organisms in California.

Since the publication in 1988 of *Lichens of California* (see Bibliography for citations), there have been many changes in lichen taxonomy and many new species recognized for the state. The most recent checklist, compiled by Shirley C. Tucker and Bruce D. Ryan in 2006, lists 1,575 taxa (including 65 subspecies, varieties, and forms), and the number continues to grow with no end in sight. *Lichens of North America* (2001) includes many California species, and the 2007 publication of *Lichen Flora of the Greater Sonoran Desert Region,* as well as numerous journal articles, have greatly enlarged our understanding of lichens in California. These sources are too bulky for most people to take into the field, however, and many have felt the need for a new guidebook, one that would reflect recent name changes, have good color illustrations, and include more crustose species.

The present book includes most of the "macrolichens" (larger lichens, not crusts) that one is likely to encounter in the state, plus a substantial sampling of the more common and distinctive crustose species. I have not provided keys or extensive descriptions of microscopic characteristics, choosing instead to use the space to include more species. Readers who wish to confidently identify most of the crustose lichens will need to consult primary sources, starting with *Lichens of North America* and the *Lichen Flora of the Greater Sonoran Desert Region*. Updated and expanded keys for *Lichens of North America* are being prepared by Irwin Brodo and will be published by Yale University Press; they will include most of the common or conspicuous lichens of California. Lichens evolve slowly, but lichen nomenclature is changing rapidly as scientists continue to unravel evolutionary relationships; scientific journals and online resources will enable the reader to keep up.

In addition to their decorative contributions to the visual environment, lichens have important roles in many ecosystems. They provide food, nesting material, and camouflage for a wide variety of wildlife. They are also a significant source of fixed atmospheric nitrogen, a critical resource for plants, in the temperate rainforests of the Pacific Northwest and in many desert environments. In arid parts of the American West they are key elements in preventing soil erosion, forming microbiotic soil crusts wherever livestock grazing or other impacts have not destroyed them.

Lichens are used by scientists and resource managers in a number of ways. They are sensitive indicators of air quality and are studied by scientists responsible for monitoring air pollution, especially by the USDA Forest Service. Certain species grow mostly on the bark of ancient trees, and have been used as indicators of old-growth forest stands. Some rare species of lichens have been factors in the reconsideration of planned timber sales in California.

The images in this book are drawn mostly from those we took for *Lichens of North America,* and in some cases I have had to reuse a photo that appeared in that volume. Those photos were all taken as 35 mm slides. Since then, after Sylvia passed away, I added more California photos as digital originals, both on my own and during trips taken with Mariette Cole and others. Most recently, with the assistance of Kerry Knudsen, I took additional photos specifically for this book, resulting in about seventy new images that appear in these pages.

For me, lichens are an essential part of the landscape. I began to notice them in 1973, when Sylvia Sharnoff and I started to photograph them as a hobby. I can't remember anymore what I thought all those patches of color on trees and rocks were, before I had ever heard the word "lichen." Without a name or a concept they just seemed part of the texture of things. But learning to recognize them as distinct life forms, and beginning to distinguish one from the other, was a revelation, a powerful invitation to find out more. I invite the reader to go on a similar journey.

This book would never have been possible without enormous assistance and encouragement from many people. Without Sylvia's initiative, it is unlikely I would have become involved with lichens at all, and many of the images in this book are hers. Starting in the late 1980s we worked as a team with Dr. Irwin M. Brodo, of the Canadian Museum of Nature, on *Lichens of North America*. Ernie's deep knowledge and guidance were absolutely central to our involvement with lichens, and the information in *Lichens of North America* has been my primary source. He corrected innumerable errors in the draft text for this book and made many suggestions for changes that improved the manuscript; my debt to him is greater than I can possibly express. I am enormously indebted as well to Kerry Knudsen, Curator of the Lichen Herbarium at the University of California, Riverside, who helped the project along from its beginning to its final stages, identifying specimens from recent field trips and answering endless questions with patience and skill; without his help, and his expertise about California lichens, the book would have been much impoverished and much less accurate.

Without the encouragement and advice of Peter H. Raven, President Emeritus of the Missouri Botanical Garden, I never would have begun the project at all, and the sponsorship of the Garden for fundraising efforts made it possible to complete the photography and writing. Mariette Cole, coauthor of *Lichens of California,* played a special role, collaborating with me in its earlier stages, working with me in the field and identifying specimens.

Heartfelt thanks also to Doris Baltzo, Shelly Benson, Curtis Bjork, Cheris Bratt, Mona Burrell, Tom Carlberg, Janet Doell, Theodore Esslinger, Phyllis Faber, Linda Geiser, Trevor Goward, Bill Hill, the late Greg Hind, James Lendemer, Bruce McCune, Brent D. Mishler, Thomas H. Nash III, Eric Peterson, Jerry Powell, the late Judy and Ron Robertson, Roger Rosentreter, the late Isabel Tavares, Shirley Tucker, John Villella, and the late Darrell Wright, all of whom provided guidance, identifications, places to hunt lichens, inspiration, and much good advice.

A special thanks goes to my editor, Jean Thomson Black, whose support and vision have been at the heart of the entire effort, and to my wife, Suzanne Guerlac, whose patience, love, and encouragement have made it a pleasure to complete this long journey.

I also wish to acknowledge and thank the following foundations for their generous support of the project:

The John Simon Guggenheim Memorial Foundation
The Hind Foundation
The Jiji Foundation
The Mead Foundation
and anonymous donors

A Field Guide to California Lichens

Introduction

- - - - - - - - - - -

Textures

If you take a walk in a place where the air is clean, a place with trees, bushes, and rocks, the chances are good there will be lichens all around you. If it has rained recently, the lichens will stand out more brightly, their colors more vivid. There might be round green patches on the rocks and tree bark, along with splashes of other colors: orange, bright yellow, gray, and brown, or the smaller twigs of bushes might have shrubby clumps of pale green, or perhaps there will be long strands of green hanging from the trees. These are not plants, and not moss (although they often grow alongside or on moss), but lichens, specialized fungi that have become "lichenized," teaming up with a photosynthetic partner that lives inside them. This partner, called a photobiont, is most often a species of alga, but in some lichens it is a colony of cyanobacteria, and when the fungus (termed the mycobiont) becomes lichenized, it forms an entirely new structure called a thallus. Lichens are classic examples of symbiosis, but most scientists consider the fungus to be parasitic on the photobiont, since it lives on the carbohydrates the photobiont produces and fungal tissue makes up most of the lichen. The fungus, however, doesn't kill its photobiont but keeps it protected and productive for many years, rather like a skilled farmer tending a crop.

There are approximately fifteen hundred species of lichens in California, each one a unique species of fungus, the dominant partner. Because the fungi found in lichens are never found free-living, scientists consider the relationship to be an obligate symbiosis. All of these fungal species have formed lichens with a few hundred species of algae and cyanobacteria, so the same photobionts are found in many different lichens.

Fungi are not plants but organisms with an ancient lineage all their own. The partnerships they form with algae and cyanobacteria are strangely analogous to those found in coral, in which tiny, colonial animals live in symbiosis with algae. Even the shapes of lichens seem to resemble those of coral more than they do of plants.

In early spring, the lichens on these oak trees in Carmel Valley are almost glowing. There are several species of *Usnea* and *Ramalina* on the branches, including lace lichen, *R. menziesii*.

California is undoubtedly the most broadly diverse state in the country, a meeting place of three large biomes: the moist coastal forests of the Pacific Northwest, the Great Basin sagebrush country, and the Mojave and Sonoran Deserts. It has a number of unique plant communities, different from anything else in the world: redwood forests, coast and inland chaparral, and foothill oak woodland, as well as a remarkable variety of mountain ranges, from the northern Klamath Ranges to the Sierra Nevada and the many southern ranges spanning coast to desert. This complexity of terrain and climate is reflected in the enormous diversity of both plants and lichens. Many of the species in this book occur only in the northwestern part of the state, where the climate and vegetation are similar to western Oregon. Other species are characteristic of the Sonoran or Mojave Desert, some are typical of the Great Basin, and many are unique to the Mediterranean climate of the central coast region, which has almost no summer rain but abundant fog. Quite a few of these "fog desert" species are included in the book, even though they are uncommon and restricted in range, because they are found only on the West Coast, some of them growing in Baja California as well.

Lichens live on sunlight, air, and water. They don't have roots to absorb nutrients, so unlike plants, they aren't dependent on soil for sustenance. This allows lichens to grow on a wide variety of substrates, although most prefer surfaces that are stable, such as rock or oak bark. Hard, dead wood (in this book called simply wood) supports many lichens that can also be found on rock, whereas soft, decaying wood provides a substrate for lichens,

(above)
In the Sierra foot-hills, boulders often have mixtures of brightly colored lichens, including orange *Caloplaca ignea* and yellow *Acarospora socialis.*

(left)
Coastal rocks in Marin County are covered with fog-zone lichens, shrubby species of *Niebla,* and a variety of crusts.

especially in the genus *Cladonia,* that also grow on soil. Redwood trees usually don't have many lichens on their lower trunks because the bark is too shaggy (although one lichen species, *Calicium sequoiae,* has so far been found only on old-growth redwoods) and most *Eucalyptus* peels too much for lichens to become established. On the other hand, I've seen lichens flourishing on old paint, rusty metal, bone, glass, plastic, old shoes, and the fabric on the tops of cars that have been parked in the same spot for years. Some grow on moss or other lichens, on leaves or conifer needles, even on living insects.

As a glance through this book will show, lichens come in many colors and have many sorts of structures. The ones with rosettes of flattened or raised lobes are generally called foliose, or leaflike. Most flat patches, usually smaller, on bark or rock are termed crustose; some of these have tiny lobes and can resemble foliose lichens, but if you can't remove the thallus without taking some of the substrate with it, then it's crustose. Many species of lichens are shrubby, growing either erect or pendant, and these are called fruticose. These growth forms sometimes intergrade, but they are useful as initial descriptors, and the book is organized around them. Some species with tiny, crowded, often overlapping lobes called squamules are often referred to as squamulose; these have been grouped with the crustose species except for those in *Cladonia,* a special case. Most *Cladonia* species have a base consisting of a mass of squamules from which single or branched stalks arise, and this genus has been placed with the other fruticose genera.

Most lichens have some sort of reproductive structure visible as disks, bumps, or wavy lines on the surface. These are either sexual fruiting bodies of the fungal partner, the apothecia; perithecia or lirellae, which contain ascospores; or several types of vegetative propagules that look like tiny

Candelariella aurella, a relatively pollution-tolerant species, is usually found on calcareous rock or concrete, but here it grows on a rusty steel utility cover on a Berkeley sidewalk.

A plastic part of an abandoned truck in the Sierra foothills has acquired a colony of yellow lichens, probably a species of *Candelariella*.

This boot hasn't held a foot for a long time, allowing *Cladonia* species with other lichens and some moss to become established.

outgrowths or piles of powder. All of these features are important in telling one species from another.

A word about observing lichens: most of the features described below are tiny, and seeing them clearly with the naked eye is difficult or impossible. A hand lens, or jeweler's loupe, with at least 10× magnification will make most features easy to distinguish, and a loupe is easy to take into the field. Seeing finer details requires a dissecting or compound microscope.

Internal Structures

The main body of a lichen, the thallus, consists mainly of microscopic threads of fungus, called hyphae. Among the hyphae lie the algal or cyanobacterial cells, in most lichens organized in a layer just beneath the surface. When the lichen is moist, the outermost layer of fungal cells becomes somewhat translucent, letting light through to the photosynthetic cells underneath, where they perform the chemical miracle of transforming carbon dioxide, water, and sunlight into carbohydrates. The fungal cells have the unique ability to induce the photobiont to leak carbohydrates into the surrounding tissue. In this way, the fungus can absorb the nutrients it needs to survive and grow.

A microscopic cross section through *Physcia aipolia*. On top is the fruiting body, an apothecium, with the asci lined up vertically. The ascus walls themselves are colorless, so you can't see them; the dark brown color comes from the ascospores inside, eight per ascus. White medullary tissue fills the interior, with photosynthetic cells of green algae lying mostly in a layer beneath the thin grayish cortex. Brown rhizines branch out from the lower surface.

Most lichens (the genus *Collema* is a notable exception) are stratified— that is, divided into layers. The outermost layer, called the cortex, consists of cells packed tightly together in a tough, gelatinous matrix that forms a sort of skin on the lichen's surface. In many lichens the cortex has pigments, which give the lichen its color. Below the cortex is the photobiont layer, with cells of algae or cyanobacteria enmeshed with fungal hyphae; this layer is usually some shade of green. In the middle of the thallus is a layer called the medulla; it is often the thickest part of the lichen and is usually white, but in some species, it is yellow, orange, or pink. Many foliose lichens have a lower cortex,

Physconia enteroxantha when dry. Lines of slightly greenish soredia can be seen on the edges of the lobes, whose tips are whiter due to a coating of frostlike pruina.

The same lichen when damp, greener because of the algal layer becoming more visible.

which may be white but is often black or mottled; other lichens, in particular the crustose species, lack a lower cortex, and the medulla is in direct contact with the substrate.

Many lichens deepen their color dramatically when they are wet, because the cortex becomes more transparent and the greener photobiont layer shows through. In some genera, such as *Physconia*, pale grays turn green; in others such as *Psora* or *Peltula*, browns become dark green or olive, and in lichens such as *Melanelixia glabra*, dull green becomes very bright. Species of the jelly lichen genera *Leptogium* and *Collema* swell and become greener and almost translucent when wet.

Recent research has found that lichens have a diverse community of microbiological organisms inside them, an assortment of bacteria and fungi that are different from the main symbiotic partners and whose roles within the lichen are not well understood.

External Structures

Foliose, or leaflike, lichens have a thallus with lobes that may be appressed (pressed against) the substrate or ascending, and they can be broad or narrow. Lobe width is given in many of the species descriptions; it is usually measured across the widest part of the lobe between branch point and tip. Foliose lichens have upper and lower surfaces that are clearly distinct from each other. The upper surface can have a variety of textures, from smooth to ridged or bumpy, and it often has a variety of reproductive structures, described below. The lower surface in some species of foliose lichens is bare, but in others it has hairlike structures called rhizines that help attach it to the substrate; these

A broad-lobed foliose lichen, *Flavoparmelia caperata*.

Xanthoria elegans is foliose, even though at first glance it might be taken for a crustose lichen.

The pendant fruticose lace lichen, *Ramalina menziesii,* on a flowering fruit tree, possibly plum, in the Napa Valley.

A shrubby fruticose lichen, *Usnea rubicunda,* on a twig near the coast.

are usually the same color as the lower cortex—white, brown, or black—and may be simple, tufted, or branched.

In fruticose lichens the upper cortex goes all the way around the branches or stalks, so there is no upper or lower surface. As in foliose lichens, the photobiont layer is just below the cortex, and the medulla lies deeper within. In some fruticose lichens the interior is solid, and in others, notably in the genus *Cladonia,* it is hollow, with the innermost part of the medulla forming

Tiny lobes of the squamulose species, *Psora nipponica,* crowd together on a mossy rock. Numerous black apothecia appear between the damp squamules.

Cladonia chlorophaea, like most species of that genus, has a squamulose
primary thallus with fruticose podetia that grow up from it.

a cartilaginous tube called a stereome. Lichens in the genus *Usnea* have an
elastic central cord.

Squamulose lichens, such as those in the genera *Psora* and *Peltula,* have
masses of tiny lobes, often overlapping and partly erect. The large genus
Cladonia is a special case; most species of *Cladonia* have a squamulose primary
thallus out of which grow stalks or branching structures called podetia that
are essentially fruticose.

Crustose lichens, as the name suggests, form crustlike patches on the substrate. Some species have tiny lobes around the edges that can be somewhat

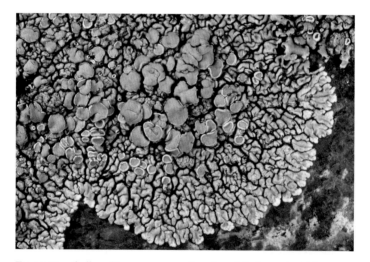

The crustose thallus of *Lecanora novomexicana* has a lobed margin, with a central area that could be described as bullate-areolate—that is, with somewhat
inflated and swollen areoles and numerous lecanorine apothecia.

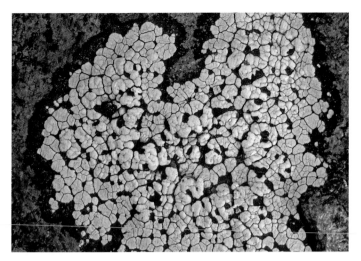

Another crustose species, *Rhizocarpon macrosporum,* has a flat areolate thallus with a black prothallus that can be seen around the margin and between the areoles. Black apothecia are slightly immersed between the areoles as well.

ascending, and some are mostly adnate (flat), but they lack a lower cortex, so the tissues of the medulla are tightly bound to the substrate. The thallus can be thick or thin, continuous, cracked (rimose), divided into irregular sections (areolate), or consist of granules or rough warts (verrucose, or verruculose when the warts are minute). In some species there is either no visible thallus or only a discoloration of the substrate, and all that shows are the reproductive structures. On rock these are called endolithic lichens; on bark they are referred to as endophloeodal. Some crustose species typically have a band of purely fungal tissue around them called a prothallus; this is usually black but

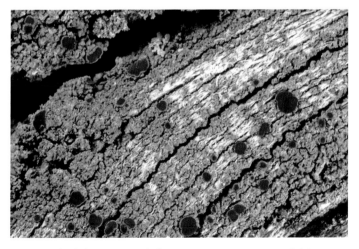

This example of the crustose *Caloplaca arizonica,* a rare species in California, has a dispersed areolate or subsquamulose thallus, appearing almost granular in the photo. On dead wood.

can be white or colored in some species. In a few genera, notably *Rhizocarpon*, a black prothallus can show between the areoles.

The color of lichen surfaces is important in telling one species from another, as is whether the surface is shiny or dull, smooth or crusty and rough (scabrose). A number of common nonreproductive features also distinguish many lichens. One of the most important is the presence or absence of pruina, a thin coating of crystals and dead cells resembling pale dust or frosting. Most often white but sometimes yellow, pruina lightens the color of the

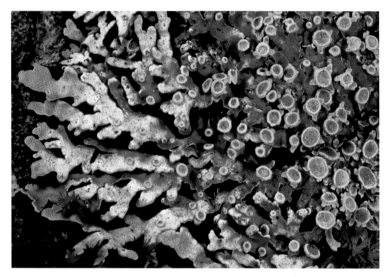

White pruina coats the lobe tips of *Physconia americana*.

surface and can be found on the lobes of a thallus or on apothecia (see Reproductive Structures, below.)

A key feature of the genus *Sticta* is the presence of cyphellae, round, white depressions on the lower surface that are actually holes in the cortex with a lining of round, thin-walled cells. Other genera have pseudocyphellae: round or irregular spots, in which a break in the cortex allows hyphae from the medulla to grow through. They are the same color as the medulla, which is usually white, and can be sunken or slightly raised, on the upper or lower surface.

Some lichens have lobules, minute lobes that develop on the thallus or along the margins of apothecia (see Reproductive Structures, below), or maculae, interruptions of the photobiont layer that appear on the thallus surface as irregular spots or reticulate blotches.

Certain lichen species have cilia, slender hairlike growths on the margins of lobes or around the rims of apothecia (see Reproductive Structures, below.) They can be unbranched or branched, pale, as in *Physcia adscendens,* or dark, as in *Heterodermia leucomela.* Unlike rhizines, they don't fasten the lichen to the substrate. Many lichens have tomentum, a cottony fuzz of very thin threads of hyphae that can grow out of a thallus, sometimes hard to see. They can grow out of the upper surface (*Peltigera rufescens*) or lower (*Lobaria, Sticta,*

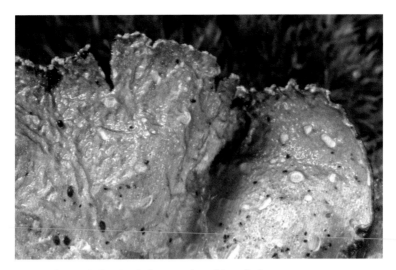

Round, white cyphellae dot the lower surface of *Sticta limbata*.

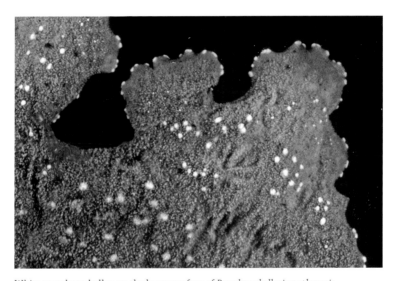

White pseudocyphellae on the lower surface of *Pseudocyphellaria anthraspis*.

Pseudocyphellaria). Others, for example, *Leptochidium albociliatum,* have cortical hairs, colorless, hyphal hairs often visible only with a hand lens.

Some lichens whose main photobiont is a green alga also contain cyanobacteria as a secondary photobiont. The cyanobacterial cells are clustered into
gall-like lumps or granules called cephalodia, most often visible as bumps on
the surface of the lichen, usually darker in color. Like algae, the cyanobacteria
carry out photosynthesis; in addition, they "fix" nitrogen—that is, transform
atmospheric nitrogen into usable chemical compounds for lichens and plants.
In this way, they assist lichens in colonizing nitrogen-poor environments and
enrich the soil for plants.

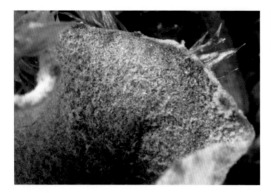

Fuzzy tomentum on the lobes of *Peltigera rufescens.*

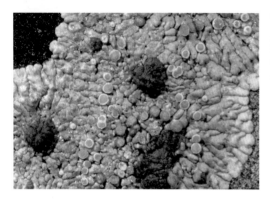

This wet thallus of *Placopsis lambii* has brown external cephalodia, pink apothecia, and green soralia.

Reproductive Structures

With few exceptions (see the Mushroom Lichens section), lichenized fungi are Ascomycetes, and their fruiting bodies are referred to as ascomata. The most common type to be seen are called apothecia, which are usually shaped like disks or cups that sit on top of the lichen's surface and have an exposed upper layer with spore-producing cells called asci. In some genera (for example, *Dermatocarpon* and *Verrucaria*) the fruiting bodies are perithecia, which are flask-shaped structures immersed in the lichen tissue and opening at the top with a small hole. There are also elongated apothecia called lirellae in some genera (*Graphis, Opegrapha*).

Apothecia share some common features, although those can be arranged in different ways. The upper surface of the disk is usually formed by the hymenium, hundreds of asci packed together vertically, interwoven with sterile hyphae that often grow above the tips of the asci and become swollen and pigmented, forming an epihymenium. Underneath the hymenium there is a layer of branched hyphae called a hypothecium. The outer margin of the apothecium is called the exciple; it may be blackened like carbon (lecideine) or pale to colorless (biatorine), or it may consist of thallus-like tissue, an amphithecium, forming a thalline margin containing algae. Apothecia with thalline margins are called lecanorine, after the genus *Lecanora;* the margin is approximately the same color as the thallus.

Numerous disk-shaped apothecia decorate the surface of *Lecanora pacifica.*

Not all apothecia follow the pattern described above. Those in the genus *Arthonia,* for example, lack an exciple and are usually irregular in shape; the elongated lirellae of graphid lichens have an exciple in two lines around the hymenium; and in the pin lichens such as *Calicium* and *Chenothecia* each fruiting body is a mass of loose ascospores, usually black or brown, called a mazaedium, enclosed by a cup-shaped exciple on top of a tiny stalk.

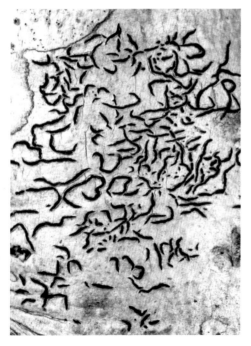

Lirellae, elongated fruiting bodies characteristic of the script lichens, resemble calligraphy on this thallus of *Graphis scripta.*

In perithecia, the other major type of fruiting body, the hymenium is very reduced and surrounded almost entirely by the excipular wall, with an opening at the top called an ostiole, through which spores are released. Perithecial exciples can be colorless or pigmented brown or black. In some species the exciple is partially or entirely enclosed by another dark layer called an involucrellum. In some lichens perithecia show as bumps on the surface of the thallus, but in others only the ostiole is visible.

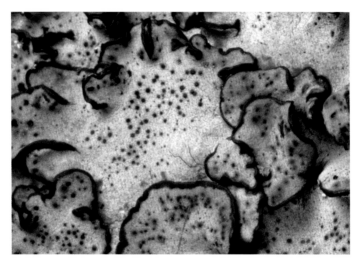

The dark spots on the surface of *Dermatocarpon miniatum* are the mouths of perithecia that are buried in the thallus.

Many lichens have tiny dots called pycnidia, usually black, but in some species pale and inconspicuous. These contain nonsexual spores called conidia that can be another way for lichens to propagate themselves. Conspicuous pycnidia can be seen near the lobe margins of, for example, species of *Tuckermannopsis*.

The finer details of reproductive structures are critical in distinguishing one genus or species from another, especially with the crustose lichens, but a more complete description of them is beyond the scope of this book. Readers who wish to learn more are encouraged to consult *Lichens of North America* and *Lichen Flora of the Greater Sonoran Desert Region* (see Bibliography) for a deeper understanding.

Of course, fungal spores by themselves cannot make a new lichen; they need to find the proper species of alga or cyanobacterium to generate a new thallus. It is a mystery how often this happens in nature, since fungus and alga come together on a microscopic scale. Perhaps it is not surprising that lichens have evolved other, nonsexual, ways of reproducing themselves by means of several kinds of vegetative propagules. These propagules are little bits of a complete lichen, and they are spread by wind and rain or by insects and other animals. They are, perhaps, the dominant way that lichens move around and become established on new substrates.

The most common of these propagules are called soredia, tiny balls consisting of a few threads of fungal hyphae surrounding a few cells of algae. Soredia do not have a cortex, and they can be fine and powdery (farinose) or coarse and grainy (granular). In many lichens the soredia are produced in aggregations called soralia; in others they are simply scattered on the surface. The types and location of soralia are important characters in identifying species: soralia may be marginal, forming along the lobe margins; laminal, occurring on the upper surface of the lobes; or labriform, like tiny lips on the lobe tips.

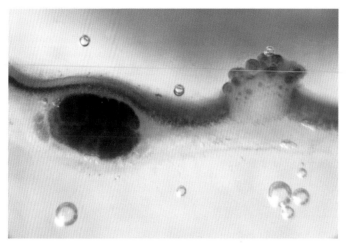

A section through *Lobaria pulmonaria*. On the right, a soralium breaks through the surface bearing round, granular soredia. The brown lump on the left, below the layer of green algae, is an internal cephalodium.

Conspicuous soralia develop along the lobes of *Ramalina farinacea*.

Many lichens have isidia, tiny outgrowths from the thallus, cylindrical or granular in shape, that have some cortex tissue. Usually they are constricted at their base, so they can break off easily and disperse to grow into a new thallus in a new location. Sometimes isidia break down or lose their cortex and become more like soredia, and the distinction between the two can become blurred. Species such as *Collema* that lack a cortex can produce granular outgrowths that are referred to as isidia, even though they are without a cortex. There are other, less common, kinds of vegetative propagules as well, such as rounded corticate thallus fragments known simply as granules, blastidia (granules that form as buds on squamules or finely divided lobe margins), and schizidia (irregular bits of stratified thallus that split off from the upper surface or form from the breakdown of small pustules).

Slightly elongated isidia on the lobes of *Platismatia herrei.*

Some lichens form a "species pair" in which one species has fruiting bodies and a sister species has only vegetative propagules but is otherwise essentially identical.

Colors and Chemistry

Many lichen fungi produce one or more special chemical compounds; there are hundreds of these products, many created by no other organisms. One of the most common is usnic acid, a yellowish pigment that gives many lichens a yellowish tinge sometimes called usnic yellow or usnic green. It is found especially in the genus *Usnea* as well as in genera such as *Flavoparmelia* and *Flavopunctelia* ("flavo" meaning yellow).

Other lichen pigments are responsible for the bright red apothecia of some *Cladonia* species and the brilliant yellows to yellow-greens of species of *Acarospora, Rhizocarpon,* and *Chrysothrix.* It is thought that these colors, often found in lichens growing in exposed, bright locations, help protect the layer

Some lichens have evolved "species pairs," with one species having fruiting bodies and another one, almost identical, having only vegetative propagules. Pictured are the fertile *Pseudocyphellaria anthraspis,* above, and its sister species, the sorediate *Pseudocyphellaria anomala,* below. Both lichens are soaking wet, making them appear much darker than if they were dry.

of algae from overexposure to sunlight. Desert lichens also tend to be more pruinose, which acts as a reflective barrier.

Some colorless chemical substances in lichens can be useful, and are sometimes essential, in telling species apart. There are some fairly simple tests that indicate their presence, and I have included "spot test" information in the species descriptions where relevant. A brief description of these tests can be found in Appendix 1.

Lichens and Their Environment

Lichens thrive in a wide range of habitats, from lush temperate rainforests to arid deserts and from seashore rocks to the tops of the highest mountains. But they don't all grow everywhere; individual species are remarkably sensitive to location. For example, rocks with moss and lichen on them are sometimes sold for landscaping, but the lichens, once moved to someone's garden, generally die. Over time, new lichens may grow on the same rocks, but the community of species will change. Lichens cannot, so far, be cultivated, one reason Irwin Brodo in *Lichens of North America* referred to them as "the essence of wildness."

In the wettest parts of California, the north coastal mountains, one finds the southernmost portion of a lichen community that extends up into British Columbia, with genera such as *Peltigera, Pseudocyphellaria,* and *Cladonia* that prefer high rainfall. In habitats of extreme rainfall, such as the Ho Valley of Olympic National Park, moss tends to outcompete the lichens, covering everything in cushions of bright green.

In the drier Coast Range forests of California, the Douglas-firs have a very diverse lichen community, often with species of *Parmotrema* and *Hypogymnia* on the branches. More montane, interior forests frequently support a spectacular growth of *Letharia,* especially on the trunks and branches of red fir; in the pine forests of the eastern parts of the state, the bright yellow *Vulpicida canadensis* stands out.

Letharia vulpina often covers conifers in montane areas; these large trunks are true firs (*Abies*) in the Trinity Alps.

Many kinds of lichens get all the moisture they need from fog and are spectacularly abundant in "fog desert" locations along the coast from Marin County south into Baja California. This unique community is dominated by species of *Niebla,* accompanied by a wide variety of lichens found nowhere else, such as *Dendrographa* and *Schizopelte,* as well as many crustose species, such as *Lecanora pinguis* and *Dirina catalinariae.* The Channel Islands, off the coast of southern California, are a refugium for such lichens, many of which used to grow on the mainland but have been eradicated by urbanization.

The oak woodlands of the Coast Range are notable for the abundance of foliose lichens such as *Flavoparmelia caperata* and *Flavopunctelia flaventior* on the trunks and larger branches of trees, with shrubby clumps of *Usnea* species covering smaller twigs in the upper canopy, and often with long strands of *Ramalina menziesii* hanging down, especially along rivers and streams.

The desert areas of the state have brilliant displays of crustose lichens on many of the rocks, such as bright orange *Caloplaca* and yellow species of

Acarospora and *Candelariella,* green *Lecanora* species, and extensive patches of brown and gray *Umbilicaria.* Rocks in the deserts and the Sierra Nevada are often mottled with gray and brown crustose lichens, such as *Lecidea atrobrunnea* and a variety of *Rhizocarpon* and *Aspicilia* species. In fact, the granite of the Sierra, if bare, is almost white; its generally gray appearance is due mostly to crustose lichens. High mountain lichens are also adapted to spending long periods under snow. When lichens are dry or in darkness, their physiological processes simply shut down, and they can wait a long time until conditions again become favorable.

Much of the black or gray color of granitic rock in the Sierra Nevada comes from crustose lichens, often growing in areas of water runoff. Photo from near Sonora Pass.

Chaparral has its own diverse lichen community, often with *Evernia prunastri,* a number of species of *Usnea,* small thalli of *Xanthomendoza,* species of *Physcia* and *Physconia,* and sometimes *Tuckermannopsis merrillii,* which often seems to grow on the older parts of manzanita. Areas of undisturbed soil can have patches of *Psora* or *Peltula* growing on them, or the bright yellow *Acarospora schleicheri.*

There are some lichen species, notably *Peltigera gowardii,* that grow mostly underwater, in streams. Others, such as *Dermatocarpon luridum,* can sometimes be semiaquatic, growing on the edge of flowing water, alternating between submerged and dry. Other lichens, such as *Hydropunctaria maura* and several species of *Verrucaria,* are typically found on the oceanic edge, in tidal zones, and many species grow only, or mostly, near the ocean.

This is only the sketchiest and most general of outlines. One of the delights

Evernia prunastri, a species of *Xanthoria,* and a species of *Candelariella* sharing a twig near Mariposa.

Species in the *Rhizocarpon geographicum* group can be semiaquatic; these thalli are growing along a seasonal stream in a subalpine meadow in the Sierra Nevada.

of hunting lichens is to discover the enormous variety of lichen communities that display unique mixtures of species. One can walk through habitats that seem as if they should support an abundance of lichens and yet there will be very few; then one comes upon a spot where the lichens are everywhere. For the most part they prefer high light levels, so very shady areas, such as the understory of older redwood forests, are lichen-poor. Lichens are often especially prominent on dead or dying trees; this isn't because the lichens killed the tree but because without leaves there's plenty of light. The traditional wisdom that moss grows on the north sides of trees applies, in a way, to lichens as well, although the preferred side isn't always north. Often it seems to be the shady, or wetter, side of the tree or rock face that has the most lichens. Species of *Lepraria,* or brilliant yellow-green *Chrysothrix,* grow most often on shaded surfaces; these are "leprose" lichens whose powdery surface can't absorb liquid water, so they especially need the moist air found in the shade of overhanging rocks or tree bases.

Most lichens are sensitive to the chemistry, especially the acidity, of their substrate. Because of this, limestone (and concrete) support a different species mix than granite, and the lichens commonly found on pine bark, which is relatively acid, are different from the ones that grow on maples, which are more basic. Some lichen species seem to be associated with certain trees; for example, in the San Francisco Bay region, California buckeyes are usually covered with bright orange *Xanthoria parietina,* and if you want to find *Lobaria, Sticta,* or *Pseudocyphellaria,* look for a bigleaf maple.

Most species of lichens are intolerant of air pollution, which is why one sees very few lichens in cities. There are a few notable exceptions, however, such as *Candelariella aurella,* which often looks like spots of yellow paint on city sidewalks. And some species of *Xanthoria* and *Candelariella* flourish in nitrogen-rich environments, like the tops of rocks where birds perch. Such lichens are also found in areas of intense agricultural activity, such as the Central Valley (otherwise quite poor in lichens), where the dust is high in nitrogen from the use of fertilizer. The ammonia-rich dust buffers the acidity of other pollutants found in these places, such as sulphuric acid. Scientists who monitor air quality have found lichens to be useful tools, since different species vary in their tolerance to pollutants, making it possible to construct maps of air quality based on patterns of lichen distribution. Since lichens are mostly very slow growing, they are heavily affected by habitat alteration and fire; abundance of lichens and high lichen diversity are good indicators that an area is relatively pristine.

Candelariella aurella on a concrete sidewalk in Berkeley. Notice how the lichen grows on the calcium-rich mortar but avoids the crystalline, siliceous quartz.

The lichens on this coastal rock thrive on the nitrogen-rich runoff from bird droppings.

Lichens have traditionally been thought of as pioneer organisms, breaking down rock into soil and establishing conditions for plants to grow. This happens through a number of paths. Many species release weak organic acids that eat away at rock, especially limestone. Many also grow inside cracks in the rock, and like roots, they expand and cause the rock to fragment. The lichens that contain cyanobacteria, either as the main photobiont or in cephalodia, release nitrogen compounds to their environment. Since nitrogen is often a limiting factor in the growth of plants, lichens in harsh environments such as deserts or mountaintops pave the way for plants to get a start, and these, in turn, accelerate the process of soil formation.

Lichen species vary widely in their growth rate. *Ramalina menziesii* can increase its weight by about a third in one year, for example, but most foliose lichens grow radially at about 0.5–4 mm per year, and crustose lichens, especially in very arid sites, grow extremely slowly. Some patches of crustose lichens are thought to be as much as four thousand years old; often the edges of a patch continue to grow outward while the center disintegrates. Lichens are found in greatest abundance and diversity in relatively undisturbed ecosystems; in literature they are often mentioned in connection with hoary, ancient forests. Some kinds, however, notably a number of species of *Cladonia* and *Peltigera*, will colonize recently disturbed areas such as clear-cuts or eroding road embankments rather quickly.

Many animals use lichens for food, nesting material, camouflage, or purposes that aren't always apparent; mites, for example, are often more abundant on lichens in trees than on the surrounding bark. Mule deer in some

A section through sandstone, with black apothecia of *Lecidea laboriosa* on top. One can see the whitish thallus under the rock surface, with a layer of green algae running through it.

Deer in the Carmel Valley have eaten all the lichens they could reach on these small oaks, creating a browse line.

parts of California have become accustomed to eating lichens such as *Ramalina menziesii,* often rearing up on their hind legs to reach it, especially in winter when other forage is scarce. Lichens are low in protein but high in carbohydrates, and they can provide needed energy for deer at their most vulnerable time of year. Because lichens are slow growing, deer can have a noticeable impact on the lichens.

Northern flying squirrels, which live in many northern California conifer forests, are known to eat lichens, especially *Bryoria* species; in parts of the Rocky Mountains they also build nests out of *Bryoria,* but I'm not aware of any reports of this behavior in California. A number of small mammals eat lichens or use them as material for nests, as do many species of birds and insects. Lichens provide excellent camouflage for nests and webs, blending them into the surroundings. Some moths have patterns of color on their wings that have almost certainly evolved to make them hard to see against lichen-covered bark.

An Anna's hummingbird (*Calypte anna*) feeds her young in a nest camouflaged with lichens.

This spider web is nicely hidden by its lichen decoration.

A moth, *Cryphia viridata,* from Mt. Tamalpais in Marin Co., is camouflaged to resemble the lichens on the trees in its environment. It is a nocturnal insect that rests during the day, so this lichen mimicry offers protection from predators. The caterpillars of this moth are reported to feed on lichens.

Lichens and People

Lichens have been used as food, medicine, padding, clothing, dye material, poison, and decoration, although records for California are scant. In southern British Columbia, native peoples boiled and ate *Bryoria* species, but it is not known if anyone in this state ate lichens. There are records of native peoples using lichens to dye porcupine quills for baskets and clothing, and the Achomawi people of northern California used *Letharia* to poison arrowheads. There are reports of *Usnea* being used for diapers; because usnic acid is an effective topical antibiotic, this practice probably helped to raise healthy babies.

Scientists have looked at lichen substances for a number of years for their possible use as medicines, and various species of *Usnea* have long been a part of many herbal compounds such as deodorants and skin creams.

USDA Forest Service scientists have an active program of tracking air quality in California by mapping patterns of lichen distribution. There are a number of online sources of information for this, for example, at http://www .fs.fed.us/pnw/pubs/pnw_gtr737.pdf.

Studying, Collecting, Conserving, and Identifying

A hand lens is the single most important tool to have with you in the field; it will make it easy to see most of the obvious features of lichens. Since lichens are so slow to grow, collecting them is not encouraged unless you are engaged in serious study. Additionally, crustose species can only be collected by removing some of the substrate, leaving a scar. On bark this is typically done with a knife or wood chisel; on rock one needs a hammer and cold chisel. Never collect in protected areas or on private land without permission. Specimens are kept in folded paper packets, not plastic bags, because they need to be kept dry or they will rot. For good information on how to collect and store specimens, see *Lichens of North America*. The California Lichen Society offers regular field trips and identification workshops for interested amateurs.

Many, perhaps most, species of "macrolichens"—not crustose, or minute —can be told apart by their general appearance and growth habit, the features one can see with a hand lens, and chemistry. With the crustose lichens it can be much more difficult. Some crustose species are distinctive, such as *Caloplaca luteominia* var. *bolanderi,* which stands out because of its unmistakable color, but most can be identified only through careful study of their microscopic characteristics, the number and type of spores, the details of ascus structure, and so forth. In the genus *Aspicilia,* for example, one usually has to examine the conidia, and these can be hard to find. Sometimes thin-layer chromatography is needed to identify the compounds in lichens more thoroughly. Even experts often have trouble naming lichens with confidence, in part because the range of variation within a species can make it resemble closely related ones. Increasingly, scientists are using molecular methods, including gene sequencing, and cladistics, a statistical method of revealing evolutionary relationships, to define genera and species.

None of this should discourage readers who want to try their hand at serious identification, however, and there are good sources of help beyond this book. Anyone who wants to dig deeper will need to have a copy of *Lichens of North America* and *Lichen Flora of the Greater Sonoran Desert Region* on hand, also *Macrolichens of the Pacific Northwest* for the northwest species, a good dissecting and compound microscope, and a few easily obtainable chemical reagents. If there is a nearby chapter of the California Lichen Society, one can participate in field trips and identification workshops where the necessary equipment is available. The Bibliography lists some of the important books about lichens, scientific journals that regularly publish articles on them, and a number of useful online resources.

As with all groups of organisms, some lichens are common and others are rare. The California Lichen Society is working with the California Native Plant

Society to identify species that are extremely uncommon or endangered. A number of different categories are used, depending on the degree of rarity and whether a species is rare globally or only in the state.

A caveat about this book is important to emphasize. It provides photos and short descriptions of most of the macrolichen species in California, along with a substantial sampling of the crusts, and under the accounts of genera, I have listed a few of the more common species not illustrated with photos. These lists are not complete, however; they leave out species that either have been found very infrequently or, occasionally, even some that might be common in limited areas. A serious observer will undoubtedly find lichens that are not mentioned herein. I would also encourage the reader to consult the keys to lichens that can be found in other books. Although I have not included them, they can be the quickest route to identification.

A number of fungi, not included in this book, look like lichens but do not have a symbiotic relationship with a photobiont and thus are not considered "lichenized." One of these is *Lichenothelia scopularia*, a dark gray perithecial crust on rock that could be mistaken for a species of *Verrucaria* or *Staurothele*. Another is *Arthopyrenia plumbaria*, with a crustose, pale gray or greenish thallus and black perithecia, found occasionally on bark in central and southern coastal areas; sectioning will reveal the absence of an algal layer. And one black crust I saw on tidal zone rock in Marin County looked like a species of *Verrucaria* but turned out to consist only of algae, without any fungal hyphae.

Our present knowledge of the lichens in California is still quite incomplete. Furthermore, the true evolutionary relationships among species of lichen fungi are just beginning to be worked out, so their classification is changing rapidly, and it is difficult for available information in print and even online to keep up. Collection data for many species are scanty, and many older identifications are questionable, so the true range of many species, especially crusts, is largely unknown; published information often reflects where collectors have gone as much as where the species actually grow. The opportunity for amateur collectors and naturalists to make a real contribution to our knowledge of California lichens, therefore, is substantial.

One of the fascinating things about lichens is that a species that is generally uncommon may be strikingly abundant in one area. I remember, for example, that Sylvia and I never saw *Ahtiana pallidula* anywhere in California until we came across a particular spot in Butterfly Valley, Plumas County, where suddenly it seemed to be on every pine tree; if this was where you lived, you might think it was a common lichen, but we never found it again. I think this element of surprise is one of the most satisfying rewards of looking for lichens, along with the fact that they grow in all of the wildest and most beautiful environments. In a world in which few habitats remain unaltered by human activities, an appreciation of lichens can strengthen the desire to preserve and defend biological diversity and protect natural areas. To study them, one must go on foot and look closely at parts of the landscape as yet untamed.

A final word on the organization of the book: I have divided the lichens into foliose, fruticose, and crustose genera. This can be a useful way for the reader to get to the right sort of lichen, but it has several difficulties. *Aspicilia,* for example, is almost entirely a crustose genus but includes one fruticose species in California, *A. californica,* and the genus *Dendrographa* now includes some crustose species that used to be in *Roccellina*. Sometimes species can produce forms that seem intermediate; nature doesn't conform to our rules! A short section also describes two kinds of mushroom lichens, species of fungus that are incompletely lichenized.

Foliose Lichens
_ _ _ _ _ _ _ _ _ _ _ _ _ _

Ahtiana

Candlewax lichens

Pale yellowish green, medium-sized foliose lichens without soredia or pseudocyphellae. Apothecia light brown, lecanorine. Lower surface pale and wrinkled with pale rhizines. Spores spherical, colorless, 1-celled, 8 per ascus. Spot tests: cortex K–, KC+ yellow, C–, P–; medulla no reactions. Uncommon on bark, mostly in drier forests of central to northern CA.

Ahtiana pallidula

Pallid candlewax lichen

Thallus pale yellow-green, loosely attached, with lobes 4–10 mm wide that can be toothed; sometimes pruinose. Apothecia tan, marginal, up to 1 cm wide, without soredia or isidia. Lower surface wrinkled with sharp ridges and pale, slender rhizines. Not common, usually on conifer twigs in northern interior montane habitats. Resembles species of _Tuckermannopsis_ in shape, but those are not yellow-green. Recent research has shown that the generic position of this species is uncertain. Photo from near Quincy, northern Sierra Nevada.

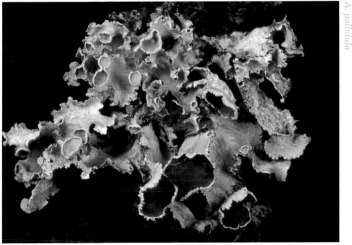

A. pallidula

Ahtiana sphaerosporella
Mountain candlewax lichen

Similar to *A. pallidula,* but with appressed, not loose, thallus and laminal apothecia. Rounded lobes are 2–4 mm wide. Lower surface pale tan, rhizines short, slender, sometimes branched, woolly. Rare on conifers in upper montane zones, central to northern CA. Compare with *Flavoparmelia caperata,* which is rarely fertile and has a black lower surface. Photo from Squaw Valley, northern Sierra Nevada.

Candelaria
Candleflame lichens

Small foliose lichens, bright to dark yellow or yellow-orange with finely divided lobes. Apothecia lecanorine. Lower surface with rhizines. Spores ellipsoid, colorless, 1-celled, 8–50 per ascus. Resembles species of *Candelariella,* but those are crustose and without rhizines.

Candelaria concolor
Candleflame lichen, Lemon lichen

Thallus small, bright yellow, forming rosettes less than 1 cm wide, with tiny overlapping, finely divided lobes that have granular margins. Usually covered with granular soredia; thallus sometimes almost entirely granular with only a few small lobes. Apothecia occasional, yellow to dark yellow or somewhat orange. Lower surface with short, white, simple rhizines. Can become quite greenish when wet. Spores 20–50 per ascus. Spot tests negative except apothecial disks sometimes K+ pink. Widespread and common on smooth tree bark, occasionally on rock, especially in nutrient-rich sites, and found in cities more than most lichens. Resembles some species of *Candelariella,* but *C. concolor* is minutely foliose, not crustose. Photo from a street tree in Berkeley.

Candelaria pacifica
Pacific candleflame lichen

Similar to *C. concolor* and recently separated from that species, but can be darker yellow or a bit more orange, with lobes that tend to be more erect and twisted. Apothecia more frequent. Lower surface partly or entirely lacking a cortex but with some white, simple rhizines. Produces slightly greenish soredia on lower surface. It also has 8 spores per ascus, whereas *C. concolor* usually has more than 30. Spot tests negative except sometimes K+ yellow-red. Fairly common on bark and wood, rarely on rock, statewide. Like *C. concolor,* resembles species of *Candelariella.* Photo from San Luis Obispo Co.

Collema

Jelly lichens

Small to medium-sized foliose lichens, dark greenish or olive to almost black; rather inconspicuous when dry, swelling and becoming gelatinous and more visible when wet. The photobiont in *Collema* is *Nostoc,* a cyanobacterium (photosynthetic bacterium), spread throughout the thallus, which lacks distinct layers or a cortex. Apothecia lecanorine. Lower surface like the upper, without rhizines, but sometimes with tufts of white tomentum. Spot tests negative; no lichen substances. *Collema* is similar in appearance to *Leptogium,* but that genus has upper and lower cellular cortices. Usually on deciduous tree bark or mossy rock, some species on soil. A few other species, not described below, occur in CA, but they are uncommon.

Collema coccophorum

Tar jelly lichen

Thallus almost black, with lobes that thicken at the tips, developing small lobules or cylindrical growths over the surface that can resemble isidia; sometimes with true, spherical isidia. Apothecia usually numerous, dark brown to black, even a bit purple, with raised margins. Spores ellipsoid, 2-celled, not constricted in the middle, 8 per ascus. On soil, especially when calcareous, or on soft sandstone, southern coast and inland, at all elevations. Photo from Organ Pipe National Monument, AZ.

C. coccophorum

Collema crispum

Scaly jelly lichen

Thallus small, black to dark green, lobate, often with overlapping (imbricate), ascending, flat or concave lobes, and with flat, scaly isidia resembling lobules; sometimes pruinose. Apothecia reddish brown, with thin margins, usually without pruina. Spores ovoid to subellipsoid with rounded ends, 4-celled,

8 per ascus. Moderately common on calcareous soil, on moss, among filamentous cyanobacteria, or on crumbling rock, southern coast and inland. Photo of a wet specimen from Banff National Park, AB.

C. crispum

Collema furfuraceum
Blistered jelly lichen

Thallus thin, membranous, medium-sized, with broad lobes 5–10 mm wide; dark olive or blackish green to almost brownish when dry, blackish green when wet. The lobes have longitudinal ridges and pustules, and abundant cylindrical isidia usually cover the ridges and central parts of the thallus. Apothecia rare. Spores narrowly fusiform to needle-shaped, 5–6-celled, often curved, 8 per ascus. *C. furfuraceum* has the largest thallus of any *Collema* in CA and is probably the most common. Can resemble isidiate specimens of *C. nigrescens,* but that typically has spherical isidia. Another isidiate, bark-dwelling species, *C. subflaccidum* (not pictured), lacks ridges and pustules and is occasional in CA. Usually on bark, especially broad-leaved trees, but also on conifers or mossy rock, statewide, coast and inland, often in drier habitats and at higher elevations.

Collema furfuraceum in the dry state.

The same thallus when damp. Photos from the Rocky Mountains, CO.

Collema nigrescens

Button jelly lichen, Bat's wing lichen

Thallus broad, lobed, with conspicuous ridges and pustules, giving it a bumpy appearance; dark olive to brownish green when dry, blackish green when wet. Usually covered with dark purplish brown apothecia, especially near the center of the thallus. Isidia usually absent; when present, spherical or sometimes club-shaped, not cylindrical as in *C. furfuraceum*. Spores needle-shaped, 6–15-celled, 8 per ascus. Fairly common on bark, especially on bigleaf maples (*Acer macrophyllum*), statewide, most often in wetter coastal areas of the Coast Range. Somewhat similar, but much rarer, is *C. subnigrescens* (not pictured), which is not isidiate but has apothecia that can sometimes be pruinose; it grows on bark in coastal habitats, San Mateo Co. to southern CA. Photo, from the Klamath Range, Trinity Co., shows a damp thallus.

Collema polycarpon
Shaly jelly lichen

Thallus blackish, forming small rosettes, with tiny radiating lobes 1–3 mm wide, thick and elongated, deeply divided, thicker at the edges, and curling up, often becoming erect. Apothecia abundant, sessile, or constricted at the base and almost stalked, dark red-brown to black, usually crowded near the center; without isidia or pruina. Lower surface slightly paler than the upper, with tufts of white tomentum (some call them rhizines, but they're typically

only a single cell thick). Spores narrowly ellipsoid, 2–4-celled, 8 per ascus, but often hard to find. Resembles *C. coccophorum,* but that species typically grows on soil and has 2-celled spores. Another similar species, *C. cristatum* (not pictured), has crisped lobes and is rare in CA. A variety of *C. polycarpon,* var. *corcyrense,* can resemble broad-lobed thalli of *C. tenax.* Uncommon on rock, especially calcareous, mostly in mountains. Photo from the Snake River Canyon, WY.

Collema tenax
Soil jelly lichen
Thallus very small, dark gray to black, foliose but often appearing almost subcrustose, sometimes forming large colonies up to 10 cm wide, isidiate to granular in the center, with lobes that vary in size and have thickened margins, often loosely attached or slightly ascending, swelling when moist. Apothecia may be abundant or absent; spherical or elongated isidia sometimes on surface. Spores fusiform, rarely submuriform and ellipsoid, mostly 4-celled, 8 per ascus, occasionally fewer. A variable species with several varieties, usually found on soil or thin soil over rock, occasionally on moss or rock, central to southern CA, especially in coastal desert areas that get periodic moisture. Another species on soil, *C. limosum* (not pictured), has a very thin, membranous thallus, almost disappearing when dry, with large apothecia up to 2 mm wide and 4-spored asci; occasional in the Coast Range and coastal habitats, Lake Co. to southern CA. Photo from Organ Pipe National Monument, AZ.

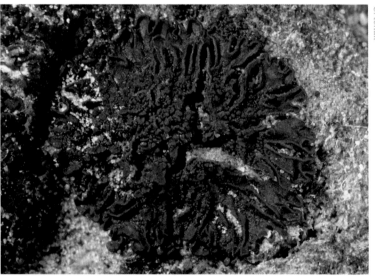

C. tenax

Dermatocarpon
Stippleback lichens
Small to medium-sized foliose lichens that grow on rock, with fairly thick lobes 1–4 cm wide, dull gray to tan, attached with a central holdfast, but

sometimes broadly squamulose with multiple attachments. Upper surface typically with tiny black or brown dots, which are embedded perithecia. Lower surface pale to dark brown, usually without rhizines, smooth or sometimes ridged. Spores ellipsoid, colorless, 1-celled, 8 per ascus. Spot tests negative; no lichen substances. *Dermatocarpon* species resemble those of *Umbilicaria*, except that those have apothecia instead of perithecia. Scattered and occasional; not encountered as frequently as *Umbilicaria*, which is often locally abundant. A few *Dermatocarpon* species besides the ones described below occur in CA. The most common is *D. leptophyllodes*, with small, single, gray lobes, 2–6 mm wide, and black perithecia and pycnidia; occasional in the central and southern coast and mountains.

Dermatocarpon americanum
American stippleback lichen
Thallus gray, umbilicate, with lobes 17–60 mm wide and black perithecia. Fairly common on all types of rock, coast and inland statewide. Many specimens previously called *D. miniatum* in the Southwest have recently been called *D. americanum* if they react red in their medulla with a rather toxic iodine solution, Meltzer's reagent (made with chloral hydrate), but not all researchers have found the reaction to be reliable. Photo from Joshua Tree National Park.

D. americanum

Dermatocarpon intestiniforme
Intestine stippleback lichen
Thallus squamulose to almost crustose, with small, convex, overlapping lobes, less than 1.5 mm wide, gray or brownish on the upper surface and pruinose. Occasional on rock, especially siliceous rock, mountain areas statewide. Photo from the Rocky Mountains, CO.

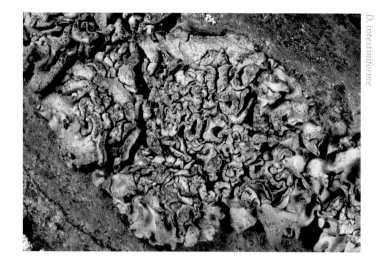

Dermatocarpon luridum

Brook lichen, Streamside stippleback

Lobes small, overlapping, mostly under 20 mm wide; pale gray to light brown when dry, but becoming bright green to gray-green when wet. Lower surface smooth or a bit papillate. When dry, resembles small, young *D. miniatum*, but grows on rock along and in streams, so it is commonly encountered in the wet state. Another semiaquatic species, *D. rivulorum*, has a veined or wrinkled lower surface; it is quite rare. Photo from Lassen National Forest.

Dermatocarpon miniatum

Common stippleback

Thallus variable, very pale tan to gray, with few lobes or many, or even curling up and becoming unattached (see vagrant form, below). Upper surface often seems pruinose; lower surface tan to black and usually smooth but sometimes a bit reticulated. This is the most common species of *Dermatocarpon;* prefers calcareous rock, found especially in the Sierra Nevada but also in coastal areas. Resembles *D. luridum* (when that is dry). Forms of *D. miniatum* with many small lobes near the center (polyphyllous) are similar to *D. intestiniforme,* but the lobes of that species are convex, and it prefers noncalcareous rocks. *Dermatocarpon moulinsii* (not pictured) is gray and single-lobed like *D. miniatum,* but with rhizines on the lower surface; uncommon in interior mountains. Photo from Coronado National Forest, AZ.

D. miniatum

Dermatocarpon miniatum vagrant form
Wandering stippleback

A fairly rare form (not officially recognized) of *D. miniatum* in which the thallus becomes unattached from the substrate and curls up, becoming almost fruticose in appearance; typically darker in color than the common form. Northern CA mountains and western slope of the northern Sierra Nevada; true range unknown. Photo from Siskiyou National Forest.

D. miniatum vagrant form

Dermatocarpon reticulatum

Sandpaper stippleback, Northwest stippleback

Thallus pale brownish gray to purplish gray, covered with light or heavy pruina. Margins often scalloped, and rounded lobes can overlap, sometimes growing into rounded clumps, even becoming vagrant. Lower surface distinctive, dark brown or black, sometimes with circular ridges, but always with papillae that give it a texture like coarse sandpaper. On rock statewide in mountains, rarely near the coast. Photo from the eastern side of the Sierra Nevada.

Esslingeriana idahoensis

Tinted rag lichen

Medium-sized foliose lichen, creamy white with a slightly yellowish tint, lobes 1.5–5 mm wide; occasionally the lobe margins have lobules. Apothecia common, brown. Lower surface wrinkled, black, with black, unbranched

rhizines. Without soredia, isidia, or pseudocyphellae. Spores ellipsoid, color-less, 1-celled, 8 per ascus. Spot tests: cortex K+ yellow, KC–, C–, P+ pale yellow; medulla K+ purplish pink on the lobe tips or margins of apothecia, KC–, C–, P–. Most closely resembles *Platismatia glauca,* which is greener, not yellowish, and typically has soredia or isidia along the lobe margins. In open conifer for-ests, northern CA mountains. Photo from Lassen National Forest.

Flavoparmelia caperata
Common greenshield
Thallus medium to large, foliose, forming rounded patches 5–20 cm wide; light yellow-green when dry, deeper green when wet. Lobes rounded, 3–8 mm wide, with clusters of granular to wartlike soredia, especially on the older, central portions of the thallus and on the raised parts of the lobes. Lower sur-face black, with black, unbranched rhizines; can be brown to naked around periphery. Medulla white. Spot tests: cortex K–, KC+ gold, C–, P–; medulla K–, KC+ pink, C–, P+ red-orange. Quite common on bark, especially on oaks, occasionally on rock; often seen on fences, particularly on slanting boards, such as roof shingles. Statewide, frequent in the central Coast Range and oak woodlands. *F. caperata* can resemble the less common *Flavopunctelia soredica,* but that species has marginal soralia and usually a few white pseudocyphel-lae. Often grows alongside *Flavopunctelia flaventior.* Photo of a damp specimen from a fence in Berkeley.

Flavopunctelia
Speckled greenshield lichens
Medium to large, yellow-green or green, foliose lichens with pseudocyphellae appearing as white spots on upper surface. Spot tests: upper cortex K+ yellow-ish, KC+ gold, C–, P–; medulla K–, KC+ red, C+ red, P–. On bark, rarely on rock.

F. caperata

Flavopunctelia flaventior
Speckled greenshield

Thallus green, foliose, appressed, with white pseudocyphellae dotting upper surface and farinose to granular soredia on the lobe surfaces and often on the lobe margins as well. Apothecia and pycnidia rare. Lower surface black with a brown edge; sparse black rhizines. Medulla white, with a continuous algal layer. Spot tests: KC+ gold from usnic acid. Fairly common on bark, often growing alongside *Flavoparmelia caperata*. On bark and sometimes rocks, statewide, except for high elevations; its range extends a bit more inland than *F. caperata*. *F. flaventior* is usually a darker, bluer shade of green than either *F. caperata* or *F. soredica,* and the white pseudocyphellae are conspicuous. Some specimens, however, can closely resemble *F. soredica.* Photo from San Luis Obispo Co.

F. flaventior

Flavopunctelia soredica
Powder-edged speckled greenshield

Similar to *F. flaventior,* except that *F. soredica* is usually yellowish green (more like *F. caperata*) with marginal soralia and sparse pseudocyphellae. Less common, mostly central and southern Coast Range. Photo from near Prescott, AZ.

Heterodermia
Fringe lichens, Centipede lichens

Small, foliose or subfruticose lichens with pale gray to whitish, narrow lobes that often widen at the tips and with conspicuous cilia; lower surface white. Spores brown, 2-celled, with unevenly thickened walls, 8 per ascus. The species found in CA are loosely attached to their substrate and have ascending lobes, appearing almost fruticose.

Heterodermia erinacea
Coastal fringe lichen

Branches white or pale bluish gray, straplike, ascending, with abundant long cilia. Apothecia black with white margins. Lower surface white and webby, lacking a cortex. Without soredia. Spot tests: medulla K+ yellow, KC–, C–, P+ yellow or P–. On shrubs and trees, central to southern coast. A similar but less common species, *H. namaquana* (not pictured), has helmet-shaped lobes with soredia on the underside; it occurs in central and southern CA coastal areas. Photo from the Channel Is.

Heterodermia leucomela

Elegant fringe lichen

Lobes white, long, narrow, ascending, linear, branching dichotomously, with marginal black cilia that sometimes branch and granular soredia on lower surface near the lobe tips. Apothecia rare. Lower surface white, cottony, sometimes pruinose, without a cortex. Spot tests: medulla K+ red, KC–, C–, P+ yellow. On hardwoods or rock, especially mossy surfaces, in coastal habitats, San Francisco Bay to Mexico. Photo from the Channel Is.

H. leucomela

Hyperphyscia adglutinata

Grainy shadow-crust lichen

Thallus small, foliose, with narrow, gray to pale green lobes 0.5–1 mm wide and rounded green soralia on upturned lobes in older, more central parts of the thallus; sometimes with black pycnidia. Lower surface pale. Spores ellipsoid, brown, 2-celled, *Physcia*-type (with unevenly thickened walls), 8 per ascus. Spot tests negative. Occasional on hardwood bark in the Sierra Nevada, Coast Range, and southern mountains. Resembles species of *Phaeophyscia* but is more closely attached, without rhizines. Another species, *Hyperphyscia confusa* (not pictured), has somewhat coarser soralia that arise on the margins and are more crescent-shaped. Photo from Sonoma Co.

Hypogymnia

Tube lichens, Bone lichens, Pillow lichens

Medium to large foliose lichens with narrow, branching lobes, usually white to pale gray or slightly bluish or brownish on upper surface. Lobes hollow

(except solid in *H. hultenii* and *H. lophyrea*), often with a hole at the tips, and can be appressed or ascending. Species with ascending lobes can appear fruticose, but they have a distinct lower surface that is black, usually wrinkled, without rhizines. Upper surface of the inside of lobes (the medullary ceiling) can be dark brown or white; this is often helpful in distinguishing the species. Apothecia usually brown, lecanorine, and raised like small urns; lobes often dotted with conspicuous black pycnidia. Spores ellipsoid, colorless, 1-celled, 8 per ascus. Mostly found on conifer bark, sometimes on shrubs, occasionally on rock or mossy soil. Compare with *Menegazzia,* which has perforations on the upper lobe surface. Can be abundant in some habitats.

Hypogymnia apinnata

Beaded tube lichen

Lobes long or short, 2–5 mm wide, white above and black beneath, tips often brown and with a hole. Usually lacking lobules. Apothecia and pycnidia frequent, brown. Medullary ceiling dark. Spot tests: cortex K+ yellow; medulla

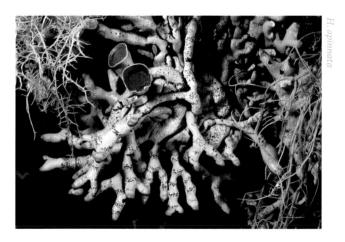

K–, KC–, P– or pale yellow. Very similar to *H. enteromorpha* but without its characteristic lobules. Occasional on conifers in fairly open forests, northern Coast Range. Note the bands of black pycnidia on the lichen in the photo. Photo from the northern Coast Range.

Hypogymnia enteromorpha
Budding tube lichen, Gut lichen
Resembles *H. apinnata,* except that the lobes vary more in shape and have a more irregular branching pattern; with frequent small lobules on the lobe margins. Spot tests: cortex K+ yellow; medulla K–, KC+ orange, C–, P+ red. One of the larger and more common *Hypogymnia* species on conifer bark and wood, mostly in coastal habitats, Monterey Co. north. When growing on wood, such as on fences near the coast, the form can be somewhat different, usually more compact. Photo from Six Rivers National Forest.

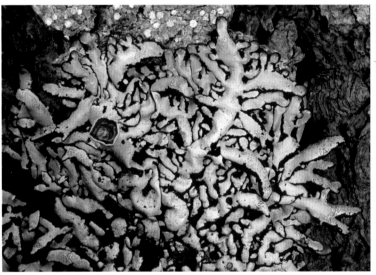

H. enteromorpha

Hypogymnia heterophylla
Seaside tube lichen
Thallus whitish, 3–10 cm long, with ascending or pendant lobes, mostly narrow but variable, with distinctly wider sections, perpendicular side lobes, and often lobules arising from the margins. Lobe tips and axils often have holes. Apothecia common, brown, on short stalks; black pycnidia also common. Medullary ceiling dark. Spot tests: cortex K+ yellow, KC–, C–, P+ pale yellow; medulla K–, KC+ red, C–, P+ red. Mostly on pine bark near the ocean, statewide, but rare in southern CA. Can resemble *H. imshaugii* and *H. inactiva,* but the perpendicular side lobes are distinctive, as are the more irregular branching and uneven lobe width. *H. imshaugii* prefers drier habitats. Photo from coastal Humboldt Co.

Hypogymnia hultenii

Powdered honeycomb lichen

Thallus small, pale yellowish gray or somewhat brownish, sometimes with a pale bluish green tint, with flat, solid, ascending lobes that have greenish soredia on the tips. Lower surface black, with tiny pits, like a honeycomb, without rhizines. Other species of *Hypogymnia*, except for the closely related

H. lophyrea, have hollow lobes. Spores spherical, colorless, 1-celled, 8 per ascus. Occasional on twigs and small branches of conifers in northwestern coastal conifer forests. Formerly classified in the genus *Cavernularia.* Photo from Southeast AK.

Hypogymnia imshaugii
Forked tube lichen
Thallus white to grayish, darker in exposed locations, with slender ascending or pendant imperforate lobes that branch in a regular dichotomous pattern. Apothecia frequent, brown, constricted at the base. Medullary ceiling white. Spot tests: cortex K+ yellow, C–, P+ pale yellow; medulla K–, KC+ pink, C–, P+ red (except for the P– chemotype). On conifer branches in relatively dry locations, especially in mountains, statewide, except for the Mojave Desert. The most common and widely distributed species of *Hypogymnia* in CA, and quite variable in form. The PD– chemotype is much like *H. inactiva,* but has the typical white medullary ceiling. An uncommon species, *H. gracilis* (not pictured), has slender, partially erect lobes, a whitish (not pure white) cavity, and numerous perforations at the lobe axils and tips; occasional along the coast, Sonoma Co. to Mexico. Photo from Castle Crags St. Park, Shasta Co.

Hypogymnia inactiva
Mottled tube lichen
Very similar to *H. imshaugii,* but the medullary ceiling is dark brown to black and it is P–. It also tends to have more black showing on the thallus as well as some perforations, and it prefers more coastal forest environments. Lobes typically erect, forking, narrower, and with more regular branching than *H. enteromorpha* or *H. heterophylla.* Occasional, central and northern CA coastal areas. Photo from Siskiyou National Forest.

H. imshaugii

H. inactiva

Hypogymnia lophyrea

Fruiting honeycomb lichen

Fertile sister species of *H. hultenii,* with flat, solid lobes lacking soredia, brown apothecia, and often a somewhat larger thallus. Lower surface black, with tiny pits, like a honeycomb, without rhizines. On twigs and small branches of conifer trees. Formerly classified in the genus *Cavernularia.* Same distribution as *H. hultenii.* Photo from southwestern OR.

H. lophyrea

Hypogymnia minilobata

Small-lobed tube lichen

Rather rare, small, up to 5 cm wide, with irregular branching; lobe tips often perforated. Without asexual propagules but apothecia common; pycnidia also common. Medulla hollow; medullary ceiling pale to dark brown. Spot tests: cortex K+ yellow, KC−, C−, P+ pale yellow; medulla K−, KC+ orange-red, C−, P−. Rather similar to *H. mollis* but lacks soredia. Found especially in old-growth coastal chaparral in San Luis Obispo and Santa Barbara Cos. and on the Channel Is. Photo from San Luis Obispo Co.

H. minilobata

Hypogymnia mollis

Grainy tube lichen

Thallus small, appressed, up to 6 cm wide, pale bluish or greenish gray, lobes soft and wrinkled, lobe tips and axils perforate, and with soredia scattered on the older, more central portions. Medullary cavity dark. Spot tests: cortex K+ yellow, KC−, C−, P+ pale yellow; medulla K−, KC+ orange-red, C−, P−. *H. mollis* (sorediate) and *H. minilobata* (fertile) form a species pair. On rock and bark,

H. mollis

sometimes on shrubs, San Luis Obispo Co. to Mexico. Photo from the Channel Is.

Hypogymnia occidentalis
Lattice tube lichen
Thallus up to 10 cm wide, with appressed lobes mostly under 4 mm wide, puffy and swollen, sometimes with black borders, smooth or wrinkled on older parts. Lobe tips usually with holes, and lobe margins sometimes have lobules, especially near the tips. Medullary cavity dark. Spot tests: cortex K+ yellow, KC–, C–, P+ pale yellow; medulla K–, KC+ orange-red, C–, P–. Mostly on bark of both conifers and hardwoods in Coast Range forests, statewide. Photo from Mt. Hood National Forest, OR.

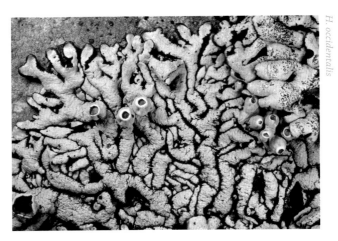

H. occidentalis

Hypogymnia physodes
Monk's hood lichen, Hooded tube lichen
Thallus variable, smaller than most *Hypogymnia* species, with pale greenish gray lobes, rather flat, appressed or ascending, but usually fanning out at the tips and with the undersides of the tips erupting into lip-shaped soralia with

H. physodes

coarse, granular soredia. Apothecia rare. Medullary ceiling usually white. Spot tests: cortex K+ yellow, KC−, C−, P+ pale yellow; medulla K−, KC+ pink or orange-red, C−, P+ orange-red. Compare with *H. tubulosa,* which has more rounded, ascending lobes, soralia that cover the lobe tips, and is P−. Occasional in the northern Coast Range. Photo from coastal WA.

Hypogymnia tubulosa
Powder-headed tube lichen
Thallus small with erect, pale bluish gray, short lobes. Upper surface of the lobe tips typically covered in powdery soredia. Apothecia very rare. Medullary ceiling white or pale brown. Spot tests: cortex K+ yellow, KC−, C−, P+ pale yellow; medulla and soralia K− or K+ yellowish to red or brown, KC+ orange-red, C−, P−. Central and northern Coast Range, occasionally inland. Compare to *H. physodes.* Photo from Marin Co.

H. tubulosa

Hypotrachyna
Loop lichens
Small or medium-sized foliose lichens, pale greenish or bluish gray, with flat lobes that have somewhat square-shaped tips and rounded axils that make the lobes overlap, leaving looplike circular spaces in between. Apothecia reddish brown, lecanorine. Without pseudocyphellae or marginal cilia, but the black lower surface has abundant, black, forked (dichotomous) rhizines that characterize the genus. Spores ellipsoid, colorless, 1-celled, 8 per ascus.

Hypotrachyna revoluta
Powdered loop lichen
Thallus small to medium-sized, pale gray, with short, irregularly shaped lobes, loosely appressed, with downward-curling (revolute) margins, tips sometimes becoming tubular. Granular soredia, darker in color, are abundant on the lobe surfaces, especially on the tips. Apothecia very rare. Rhizines sparse to abundant. Spot tests: cortex K+ yellow, KC−, P+ yellow; medulla K−,

KC+ red, C+ pink, P–. On rocks, soil, and bark, mostly in coastal southern CA, occasionally farther north. Photo from Santa Barbara Co.

H. revoluta

Hypotrachyna sinuosa
Green loop lichen

Thallus small with loose, somewhat ascending lobes usually less than 1.5 × 3 mm. Lobes distinctive pale yellowish green with powdery soredia on the older tips. Spot tests: cortex K–, KC+ gold, UV–; medulla K+ yellow turning red, KC–, C–, P+ yellow. On twigs and small branches in moist but open forests, central CA to OR, most frequent in far northern CA. Photo from the Gifford Pinchot National Forest, WA.

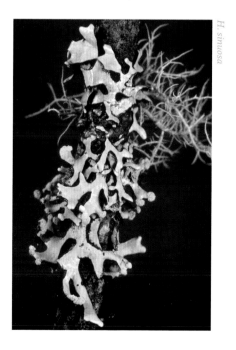

H. sinuosa

Koerberia sonomensis
Faded rock lichen

Small (up to 1 cm) foliose lichen containing the cyanobacterium *Scytonema* forming irregular, pale tan to greenish gray or olive rosettes on rock. Usually with elongate lobes at the margin and becoming lobulate in the older parts of the thallus, sometimes with reddish brown apothecia. Spores simple, narrowly ellipsoid when young, becoming 2-celled and fusiform, colorless, 8 per ascus. Spot tests negative. Uncommon and inconspicuous, mostly in mountains but can occur near the coast. A similar species, *K. biformis* (not pictured), is dark olive green with long, cylindrical isidia on upper surface; occasional in Coast Range forests and the Sierra Nevada. *K. sonomensis* can superficially resemble *Caloplaca demissa,* a crustose species with green algae as the photobiont that produces granular soredia in laminal soralia. Photo from Mt. Baker–Snoqualmie National Forest, WA.

K. sonomensis

Leptochidium albociliatum
Whiskered jelly lichen

Thallus small, foliose, dark greenish black when dry, translucent dark green when wet. Lobe margins with a fringe of stiff, short, white hairs. Apothecia, when present, have reddish brown disks, often with white hairs on the margins. Lower surface black, with a tomentum of white to brownish hairs. Spores ellipsoid, colorless, 2-celled, 8 per ascus. Spot tests negative. Resembles a species of *Leptogium,* but the marginal white hairs are distinctive, and the thallus shows a thin white medulla, which is absent in *Leptogium.* Widespread and moderately common, usually on mossy shaded rock, occasionally on soil, statewide, coast and inland.

Leptochidium albociliatum when dry. Photo from Mendocino Co.

A wet thallus of *Leptochidium albociliatum*. Photo from Trinity Co.

Leptogium
Jellyskin lichens

Small to medium-sized foliose to subfruticose jelly lichens, with dark brown or dark gray thalli, greener and translucent when wet, thin, and often rather inconspicuous; some species with brownish apothecia. Photobiont is *Nostoc*, a cyanobacterium. Frequently found growing in a tangle with moss on bark and rock. Spores colorless, 4-celled to muriform, 4–8 per ascus. Spot tests negative. Unlike *Collema*, *Leptogium* species have upper and lower cortices and are usually somewhat grayer or browner when dry. A few species of *Leptogium* other than those described below occur in CA. The two most commonly encountered are *L. plicatile*, with a dark, adnate thallus and poorly developed cortex, rather like a species of *Collema*, occasional on rock or soil in southern coastal mountains; and *L. teretiusculum*, which has small, fruticose, brown cushions with long appendages like isidia; on bark or mossy rock in the Coast Range statewide.

Leptogium californicum

California jellyskin

Thallus small, brown to greenish when dry, greenish black when wet. Wrinkled lobes are fairly flat and finely divided, 0.5–3 mm wide. Apothecia common, brown to dark brown, 0.1–1 mm wide. Lower surface pale gray to cream, smooth, with tufts of white hairs. Fairly common on mossy rocks, typically embedded in the moss, sometimes on soil or tree bases, statewide. Resembles *L. gelatinosum,* which has somewhat broader lobes and is less common, and *L. lichenoides,* which is usually smaller, with even more finely divided and more wrinkled lobes and a grayer, duller color. Photo from Lassen National Forest.

L. californicum

Leptogium gelatinosum

Petaled jellyskin

Thallus small, grayish to reddish brown, almost black when wet, lobes 1–4 mm wide and usually ascending, more or less entire, wrinkled, with margins that are often toothed or with lobules. Apothecia common, brown; not isidiate. Occasional on mossy rock or soil, sometimes on trees, statewide. Compare to

L. gelatinosum

L. californicum, above, and to the much rarer *L. polycarpum.* Photo from Klamath National Forest.

Leptogium lichenoides

Tattered jellyskin

Thallus small, forming cushions of dark brownish gray ascending and wrinkled lobes, 1–4 mm wide with finely divided margins; edges with numerous tiny cylindrical outgrowths resembling isidia. Apothecia common on lobe surfaces, red-brown, concave, 0.2–0.7 mm wide. Fairly common on mossy rocks, especially calcareous ones, also on mossy bark, statewide, except for the Mojave Desert. Compare to *L. californicum,* above. Photo from southern BC.

L. lichenoides

Leptogium palmatum

Antlered jellyskin

Thallus medium-sized, rather brownish, foliose, with erect lobes that branch and curl in at the tips, resembling tubes or antlers. Apothecia usually numerous on the lobes, red-brown. Widespread on mossy rock, sometimes soil, statewide, except for the Mojave Desert, mostly found in Coast Range. The hornlike lobe tips give it a form unlike any other *Leptogium* species, although small young specimens can resemble *L. californicum.* Photo from Six Rivers National Forest.

Leptogium platynum

Large-spored jellyskin, Wrinkled jellyskin

Thallus medium-sized, 1–7 cm wide, adnate; bluish to brownish gray when dry, dark green when wet. Lobes broad, 1–6 mm wide, minutely wrinkled, and often with minute lobules on the surface. Apothecia numerous, small (0.2–0.5 mm), on the surface or somewhat immersed. Lower surface with small tufts of white hairs. Uncommon on mossy soil or rock in northern and southern coastal mountains. Photo from Mendocino Co.

Leptogium polycarpum

Four-spored jellyskin

Thallus resembles *L. gelatinosum,* but the apothecia remain half-buried in the surface, and the asci contain only 4 spores. Rather rare on bark in northwestern forests.

Leptogium pseudofurfuraceum
Dimpled jellyskin

Thallus medium-sized, 2–8 cm wide; brownish gray when dry, greenish gray when wet. Lobes fairly broad, 3–8 mm wide, usually wrinkled and folded, and at least partly covered with cylindrical isidia. Apothecia occasional, small, brownish. Lower surface thickly white tomentose. Usually on bark, rarely on rock, statewide, typically at mid-elevations in interior and coastal mountain ranges. Somewhat resembles *L. siskyouensis* (not pictured), a smaller and grayer (less brown) species found occasionally in the northern Coast Range and Sierra Nevada, sometimes in coastal locations. Photo from Apache-Sitgreaves National Forest, AZ.

Leptogium saturninum
Bearded jellyskin

Thallus medium-sized, 2–10 cm wide, dark gray to olive, with fairly broad lobes, 3–10 mm wide, that curl up at the edges. Upper surface dull, with granular isidia, especially toward the base of the lobes. Apothecia rare, brownish. Lower surface covered with dense, short white hairs. Occasional on bark, sometimes mossy rock, in coastal mountains statewide. A similar, but rarer species, *L. burnetiae* (not pictured), has aggregates of coralloid isidia, whereas the isidia of *L. saturninum* are granular and usually scattered across the lobe surface. Photo from western slope of the Cascades, OR.

L. saturninum

Leptogium tenuissimum
Thin jellyskin lichen

Thallus small, dark gray or brownish, with minute lobes only 0.1–0.2 mm wide. Whereas most species of *Leptogium* have tissue made up of small,

L. tenuissimum

rounded cells (pseudoparenchyma) only on the cortical surface, this species is made up of it all the way through. Thallus lobes sometimes have margins fringed with tiny branches that resemble isidia. On mossy, especially calcareous, soil or rock, occasionally on bark, in central and southern CA, most frequently in coastal mountains. Somewhat resembles *L. lichenoides;* may also be confused with *Polychidium muscicola,* which differs in the spores. Photo from San Luis Obispo Co.

Lichinella nigritella
Rock licorice
Thallus small, foliose to almost fruticose, often umbilicate, with jet black, rather glossy, crowded and wavy lobes; often with spherical to scalelike isidia. Apothecia rare. Photobiont is a cyanobacterium similar to *Gloeocapsa,* giving *Lichinella* a gelatinous look like *Collema,* which contains filamentous strands of *Nostoc.* Spores simple, ellipsoid, small, colorless, 16–24 per ascus. Spot tests negative. Occasional on rock, inland and coastal, mostly in mountain areas, Mt. Shasta south. Similar to another species, *L. cribellifera* (not pictured), which has rounder, less branched lobes, lacks isidia but is usually fertile, and is rarer in CA. Another CA species, *L. stipatula,* is dwarf-fruticose, producing tiny cushions a few mm high of branched cylindrical lobes, black, with terminal apothecia; on rock in central and southern coast and montane areas. Several other species, not described here, occur in CA. Photo from Santa Barbara Co.

L. nigritella

Lobaria
Lungworts, Lung lichens
Medium to large foliose lichens, typically with broad, often squarish lobes, pale creamy brownish gray or, in some species when damp, green. Apothecia, when present, are lecanorine, brown to reddish, on the surface or along the margins. Photobiont either green algae or cyanobacteria; green algal

species have internal cephalodia containing cyanobacteria that appear as small bumps on the lower surface. Lower surface with a cortex, pale brown, usually covered with a short, fuzzy tomentum also pale brown. Spores fusiform, colorless or brownish, mostly 2–4-celled, 8 per ascus. *Pseudocyphellaria* and *Sticta* can look similar on the upper surface but have pseudocyphellae or cyphellae underneath. The three *Lobaria* species in CA all occur mainly in the northwestern part of the state, in well-lit sites in the Coast and Klamath Range forests.

Lobaria hallii
Gray lungwort

Thallus pale gray with broad lobes, dotted with grayish brown soredia on the surface and margins; the soredia often form little rings. Upper surface, especially at the lobe tips, has tiny, colorless hairs, though these can be hard to spot. Photobiont is a cyanobacterium. Spot tests negative. On bark of poplars and bigleaf maples, Mendocino Co. to OR. Shares a similar habitat with and can resemble *L. scrobiculata,* which is usually somewhat yellowish in color, lacks the colorless hairs, and differs in chemistry. Photo from Humboldt Co.

L. hallii

Lobaria pulmonaria
Lungwort, Lung lichen

Thallus large, lobed, often up to 20 cm long, with conspicuous ridges and depressions that suggest lung tissue. Color varies from pale creamy or yellowish brown when dry to bright green when wet. Soralia frequent along the ridges, with isidia developing among them. Apothecia infrequent, reddish brown. Primary photobiont is green algae, but the lower surface has cephalodia like tiny warts that contain cyanobacteria. Spot tests: medulla K+ yellow to red, KC–, C–, P+ orange; or K+ red, KC–, C–, P+ yellow. On bark, mostly of deciduous trees, sometimes on mossy rock or wood, in mature Coast Range forests,

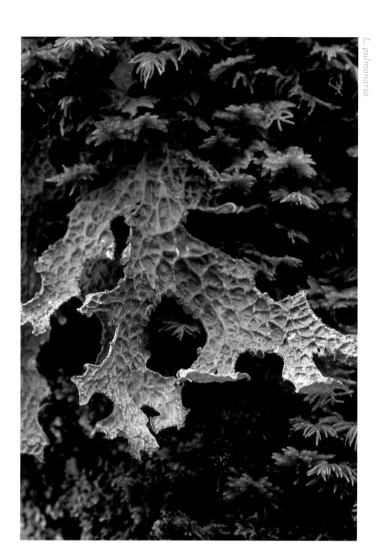

San Mateo Co. northward. The species has sometimes been used by scientists as an indicator of ancient forests. It is also a traditional source of dye. Considerably more common and widely distributed than other *Lobaria* species. Photo of a damp lichen from Six Rivers National Forest; the presence of apothecia is unusual.

Lobaria scrobiculata
Textured lungwort

Thallus creamy yellowish to pale brownish, lobes 1–2 cm wide, usually with low ridges and shallow depressions, without colorless hairs on the tips. Photobiont is a cyanobacterium. Soralia bluish gray, round or irregular, with granular soredia, on the lobe margins and surface. Spot tests: cortex K–, KC+ yellowish; medulla and soralia K+ yellow to orange, KC– or pink, C–, P+ orange.

On broadleaved trees, especially bigleaf maples, occasional on conifers, wood, or mossy rock, in northern coastal or moist inland forests. Photo from the Gifford Pinchot National Forest, WA.

Massalongia carnosa
Rockmoss rosette lichen

Small, brown, foliose lichen, more olive-colored when wet, with numerous narrow lobes, 0.5–1.5 mm wide, often overlapping and becoming almost squamulose; margins with isidia or tiny lobules. Photobiont is the cyanobacterium *Nostoc*, forming clumps in a layer inside the thallus. Apothecia reddish brown, with lighter biatorine margins. Spores colorless, 2-celled, 8 per ascus. Spot tests negative. Occasional on mossy rock, sea level to mid-elevation, in Coast Range and inland mountains statewide. Resembles *Fuscopannaria* or *Pannaria*, but lacks a hypothallus and the spores are different. Photo from Southeast AK.

Melanelia

Camouflage lichens

Small to medium-sized foliose lichens, usually a dark greenish brown color that blends into the background and makes them hard to see. Generally greener when wet. Usually closely attached, some species with lecanorine apothecia, frequently with pseudocyphellae on the surface. Lower surface tan to black with unbranched rhizines. Spores ellipsoid to spherical, colorless, 1-celled, 8–32 per ascus. On bark and rock. Species in the genera *Melanelixia* and *Masonhalea,* below, used to be classified within *Melanelia* but were separated on the basis of molecular studies. The rock-dwelling *Xanthoparmelia loxodes* and *X. subhosseana,* formerly in the genus *Neofuscelia,* strongly resemble *Melanelia* (in the broad sense) as well. Camouflage lichens can be abundant in some locales, such as on oak trees in open forests on the western slope of the Sierra Nevada.

Melanelia disjuncta

Mealy camouflage lichen

Thallus very dark brown to blackish olive, usually shiny, especially at the periphery, with lobes 0.8–1.5 mm wide, convex in the older, central parts but flattening at the edges. Middle of the thallus typically covered with dark, coarse, granular soredia that resemble isidia; often with small pseudocyphellae near the lobe margins. Apothecia rare. Lower surface dark brown to black. Spot tests negative, but UV+ white. On rock, especially granite, rarely on bark or wood, statewide, especially in inland mountains such as the Sierra Nevada. Resembles *M. tominii,* which has more conspicuous pseudocyphellae and a medulla that reacts C+ red. Photo from southern WA.

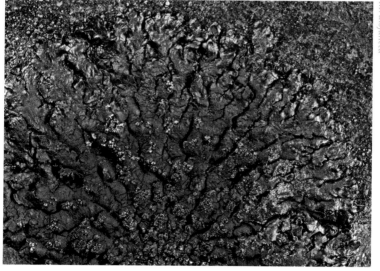

M. disjuncta

Melanelia tominii

Dimpled camouflage lichen

Thallus rather shiny, yellowish or olive to almost black, sparsely sorediate in both central area and margins; narrow lobes 1–3 mm wide, usually with conspicuous pale to dark pseuocyphellae. Apothecia frequent, sessile, up to 6 mm wide. Lower surface dark brown to black, often paler at the margin. Spot tests: medulla K–, KC+ red, C+ red, P–. On rock statewide, mostly in interior mountain ranges. Compare with *M. disjuncta,* above. Photo from the Mojave Desert.

M. tominii

Melanelixia

(See *Melanelia,* above.)

Melanelixia californica

California camouflage lichen

Thallus olive to bluish green when dry, bright green when wet, wrinkled toward the center, usually with brown, concave apothecia 1–8 mm wide that are constricted at the base. Without soredia or isidia, but sometimes with minute hairs on the margins of lobes and apothecia. Lower surface dark brown to black, often paler at the margin. Spot tests: medulla K–, KC+ red, C+ red, P–. Mainly on bark, especially of deciduous trees, occasionally on rock, statewide, but most often in the Sierra Nevada and southern mountains, less frequently along the coast. Resembles *Melanohalea multispora* and *M. subolivacea,* which have thinner thalli and are C–. Photo from Mt. Palomar, San Diego Co.

M. californica

Melanelixia fuliginosa
Shiny camouflage lichen

Thallus thin, mostly appressed, dark olive-green to brown and shiny, with lobes 1–3 mm wide, usually wrinkled and pitted, with cylindrical, often branched isidia up to 0.8 mm long. Without pseudocyphellae, but apothecia frequent, up to 6 mm across. Lower surface dark brown to black, often paler at the margin. Spot tests: medulla K– or K+ violet in orange pigmented spots, KC+ red, C+ red, P–. Occasional on bark, rock, or wood, mostly in Coast Range

M. fuliginosa

oak forests. Resembles *M. subaurifera,* which is less shiny, has much finer, unbranched isidia and often soredia as well, and differs in chemistry. Photo from coastal WA.

Melanelixia subargentifera
Whiskered camouflage lichen
Thallus olive to brown, with a rough, pitted surface; often pruinose or with minute hairs; the lobes often lift slightly at their margins. Soralia round, gray-brown, with coarse, granular soredia on the lobes and especially on the margins, crowded together in the central portion of the thallus. Lower surface dark brown to black, often paler at the margin. Spot tests: medulla K–, KC+ red, C+ red, P–. On bark, wood, and mossy rock, statewide, especially in mountains. Photo from the Rocky Mountains, CO.

M. subargentifera

Melanelixia subaurifera
Abraded camouflage lichen
Olive to brown thallus, shiny or dull, with rounded lobes 1–6 mm wide. Usually with a mixture of soredia and isidia; the isidia break down into granular soredia, leaving yellowish patches when rubbed off. Apothecia rare; pseudocyphellae absent or inconspicuous. Lower surface pale to dark brown or black, often paler at the margin, moderately rhizinate. Spot tests: medulla K–, KC+ red, C+ red, P–. Occasional on bark or wood, rarely on rock, statewide, mostly in mountain areas. When the thallus is mostly isidiate, it can resemble *M. fuliginosa.* Photo from southern ON.

Melanohalea
(See *Melanelia,* above.)

Melanohalea elegantula
Elegant camouflage lichen
Thallus small to medium-sized, lobes 1–7 mm wide, surface olive to brown and usually dull, often whitish gray with pruina. Usually isidiate, with conical or

M. subaurifera

cylindrical isidia, that are sometimes globular but never flat or spoon-shaped. Isidia arise as warts that often have pseudocyphellae at their tips. Apothecia uncommon. Lower surface tan to black. Spot tests negative. Mostly on bark, occasionally on mossy rock, statewide, rather common in inland mountains. Resembles *M. exasperatula,* which is shinier, with hollow and somewhat flattened isidia, and lacks pseudocyphellae; also resembles *M. subelegantula,* with slightly flattened, not hollow, isidia. Photo from Modoc National Forest.

M. elegantula

Melanohalea exasperatula

Lustrous camouflage lichen

Thallus olive-green and shiny, very appressed with a rough (*exasperatis* in Latin) surface. Isidia are constricted at the base and hollow, changing from rounded to flattened as they develop, becoming almost like lobules. Apothecia uncommon; without soredia or pseudocyphellae. Lower surface tan to dark brown. Spot tests negative. On wood or bark, rarely on rock, mostly in central to northern interior mountains. The shape of the isidia distinguishes it from *M. elegantula,* the much more common species of the two; compare also to *M. subelegantula.* Photo from Flathead National Forest, MT.

M. exasperatula

Melanohalea multispora

Many-spored camouflage lichen

Thallus quite brownish, with lobes 1–3 mm wide. Apothecia common; without soredia or isidia, although often with small rounded lobules on older parts of the thallus. Pseudocyphellae, if present, develop on the lobe margins. Lower surface dark brown or black. Spores 12–32 per ascus. Spot tests negative. Occasional on bark, especially on deciduous trees or on shrubs, statewide. Very similar to *M. subolivacea,* but with more spores per ascus. Also somewhat resembles *Melanelixia californica,* which is typically greener. Photo from Shasta-Trinity National Forest.

Melanohalea subelegantula

Wavy camouflage lichen

Thallus olive to green with wavy rounded lobes that are often pruinose; isidia not hollow, initially cylindrical but becoming flattened. Spot tests negative.

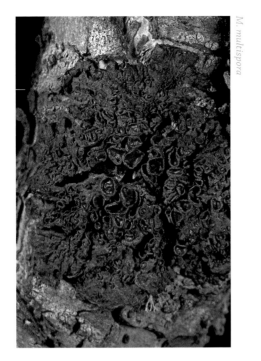

Uncommon on bark, never on rock, mostly in the Sierra Nevada. Similar in appearance to *M. exasperatula,* except for the structure of the isidia and the pruinose lobes. Photo from western ID.

Melanohalea subolivacea
Brown-eyed camouflage lichen

Thallus small to medium-sized, thin and appressed, but lobes tips sometimes raised, olive to brown, dull or shiny. Apothecia common, dark brown. No soredia or isidia, sometimes with a few pseudocyphellae. Spores 8 per ascus. Spot tests negative. Fairly common on bark, especially of broad-leaved trees, rarely on wood, statewide, in drier habitats of all mountain ranges. Resembles *M. multispora,* but more common, and *Melanelixia californica,* which is C+ red. Photo from northern AZ.

Menegazzia subsimilis
Treeflute

Thallus small to medium-sized, foliose, with narrow lobes, green at the edges becoming almost white in the center. Lobes hollow, often brown at the tips, with scattered, round perforations. Collar-shaped soralia with coarse granular soredia are usually found on short lobes on the surface; the soralia tend to fragment in older thalli. Spot tests: cortex K+ yellow, KC–, C–, P–; medulla K+ yellow, KC–, C–, P+ orange. A Pacific Northwest species. On bark, especially alder, in northwestern coastal environments. A similar, but rarer species, *M. terebrata,* has soralia that form round, smooth, powdery caps with fine soredia. Photo from coastal northern WA.

Nephroma
Kidney lichens, Paw lichens

Medium-sized, occasionally large, foliose lichens, tan to dark gray-brown, with kidney-shaped apothecia on the *lower* surface of lobe tips. Lobes often curl up at the edges, making the apothecia visible. Lower surface smooth or may have a thin tomentum, without rhizines. Photobiont is the cyanobacterium *Nostoc* for the species described below. Spores subfusiform to fusiform, pale brown, septate, thin-walled, 8 per ascus. Spot tests mostly negative.

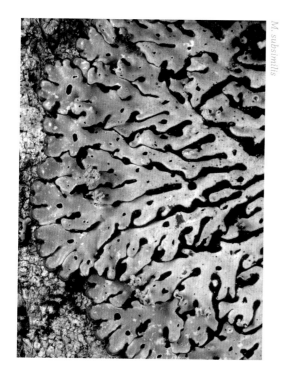

Nephroma species are somewhat related to those of *Peltigera* but have a lower cortex (lacking in *Peltigera*), and only *Nephroma* has apothecia on the lower surface.

Nephroma helveticum
Fringed kidney lichen
Thallus tan to dark brown, with lobes up to 5 mm wide. The lobes have fringes of tiny lobules and flattened isidia, which also can grow on the upper surface.

Apothecia usually present, brown, with lobulate margins. Lower surface dark brown, smooth or with fine hairs. Spot tests mostly negative, but sometimes medulla reacts K+ pale yellow. On bark and mossy rock in conifer forests, central and northern CA, Coast Range and Sierra Nevada. Photo shows a damp lichen from the Gifford Pinchot National Forest, WA.

Nephroma laevigatum
Mustard kidney lichen

Thallus thin and closely appressed, with lobes 3–9 mm wide, dark brown, often with lobules on the margin. Apothecia frequent. Lower surface tan, smooth. Medulla dark yellow. Spot tests: medulla reacts K+ reddish purple. Occasional on trees and rocks in coastal forests, San Francisco Bay to OR. Photo from Vancouver Is., BC.

N. laevigatum

Nephroma parile
Powdery kidney lichen

Thallus medium to large, usually dull dark brown or gray, but can turn olive when wet. Lobes 4–8 mm wide with gray or brownish soredia along the margins and often on the lobe surface. Apothecia rare. Lower surface smooth, tan. Fairly common on bark and mossy rocks in shady forest sites statewide, but most frequent in far northern CA. At first glance, can be mistaken for *Peltigera collina,* which also has marginal soredia, but that species has a pale, somewhat veined and webby lower surface with pale rhizines. Compare also with *Sticta limbata,* which has round cyphellae on the lower surface. *N. parile* is the only species of *Nephroma* with marginal soredia. Photo from MT.

Nephroma resupinatum

Pimpled kidney lichen

Thallus gray to dull brown. Lobes thick, ascending, forming rosettes 3–10 cm wide. Apothecia common, brown. Lower surface with a fuzz of tomentum and pale, bare, pimplelike bumps. Occasional on bark and mossy rock in humid localities, northern coast and inland forests. Resembles *N. helveticum,* except for the bumps on the lower surface; also, the backs of the apothecia in *N. resupinatum* are often hairy, whereas those of *N. helveticum* are wrinkled and rough. Photo from Wenatchee National Forest, WA.

Pannaria rubiginosa
Brown-eyed shingle lichen

Thallus small to medium-sized, foliose; light bluish gray to slightly tan when dry, dark gray when wet (as in the photo). Lobes somewhat ascending, 0.7–4 mm wide. Apothecia common, reddish brown, set off by distinct lecanorine margins. Sometimes with a visible, blue-black hypothallus. Lower surface pale, covered with a dark bluish tomentum. Spores ellipsoid, colorless, 1-celled, 20–24 × 10–12 µm, 8 per ascus. Spot tests: medulla K–, KC–, C–, P+ orange but often hard to detect, or P–. Compare with *Fuscopannaria leucostictoides,* which has much smaller, squamulose lobes, P– medulla, and often has a conspicuous black prothallus; also resembles *P. rubiginella* (not pictured), a smaller-lobed and smaller-spored, rare species from central CA. Occasional on bark, usually in shady habitats, central Coast Range. Photo from Monterey Co.

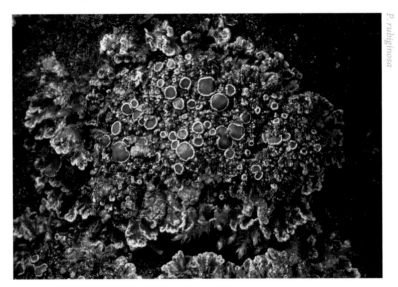

Parmelia
Shield lichens

Medium to large foliose lichens with lobes that are somewhat square-shaped at the tips, pale bluish gray, or often light brown in exposed locations, lacking cilia. Upper surface usually with a network of ridges and depressions and white pseudocyphellae or maculae. Lower surface black, with rhizines. Apothecia lecanorine. Spores ellipsoid, colorless, 1-celled, 8 per ascus. On bark, wood, and rock.

Parmelia hygrophila
Western shield lichen

Thallus blue-gray to greenish, with a network of pseudocyphellae that extends to the margins and dull isidia that can resemble granular soredia. Apothecia rare. Rhizines either unbranched or sometimes forked. Spot tests:

upper cortex K+ yellow, C–, KC–, P+ yellow; medulla K+ yellow becoming red, KC–, C–, P+ yellow to orange. On bark, most often in the wet, oceanic habitats of the Pacific Northwest, but also on the western slope of the Sierra Nevada and occasionally in southern mountains. Photo of a damp specimen from the eastern side of the Cascades, OR.

Parmelia saxatilis

Salted shield lichen, Crottle

Thallus adnate, greenish or bluish gray but often quite brown in exposed habitats. Lobes 2–4 mm wide, with ridges, depressions, a network of pseudo-cyphellae, and cylindrical or slightly branched isidia that form mostly on the ridges. Apothecia with isidiate margins. Rhizines thick, unbranched or

forked. Spot tests: medulla K+ yellow becoming red, KC–, C–, P+ yellow to orange. On rock, occasionally on wood, rarely on bark, especially in mountains, coast and inland, statewide, except for the Mojave Desert. *P. saxatilis* may have been used as a medicinal tea by the Maidu people of southern CA. Compare with *P. squarrosa.* Another, much rarer, species is *P. pseudosulcata* (not pictured); it has narrow, dichotomously branched lobes with weak ridges and reacts K–, P+ red-orange. Photo from Southeast AK.

Parmelia squarrosa

Thallus pale bluish gray, usually whiter toward the center. Lobes 1–5 mm wide, ridged, with pseudocyphellae, and with cylindrical isidia that can become dull and almost squamulose, most abundant on the margins. Apothecia not uncommon. Rhizines unbranched and slender when young, soon becoming squarrose, with short, perpendicular side branches. Spot tests: medulla K+ yellow becoming blood red, KC–, C–, P+ yellow to orange. Resembles *P. saxatilis,* but that has rhizines that are thick and unbranched or sometimes forked. Uncommon on bark or mossy rock, only in far northwestern CA. Photo from coastal OR.

P. squarrosa

Parmelia sulcata
Hammered shield lichen

Thallus pale bluish gray, sometimes browning at the edges, or completely brown in sunny locations. Lobes 2–5 mm wide with ridges, depressions, and whitish pseudocyphellae; ridges and cracked lobe margins often with powdery soredia. Apothecia rare. Rhizines slender and unbranched when young, becoming dense and squarrose with age. Spot tests: medulla K+ yellow becoming blood red, KC–, C–, P+ yellow to orange. Widespread, common, and quite variable in appearance; frequent on bark and wood in open habitats,

P. sulcata

occasionally on mossy rock or soil, coast and inland, at low elevations and in mountains, statewide, except for the Mojave Desert. A similar species, *Parmelia barrenoae* (not pictured), has simple, not squarrose rhizines and soralia that are usually less abundant and rather eroded; fairly common in some habitats, such as in Yosemite National Park. Photo from BC.

Parmelina coleae
Fringed shield lichen

Thallus medium-sized, foliose, white to pale bluish or yellowish gray, closely appressed. Lobes 2–3 mm wide and often divided dichotomously, without pseudocyphellae, soredia, or isidia. Apothecia lecanorine, dark brown. Lobe margins sometimes with black cilia, 0.2–1 mm long. Lower surface black, with long, unbranched rhizines visible at the margins; along with the cilia they give the lichen a fringe around the edges. Spot tests: cortex K+ yellow; medulla K–, KC+ red, C+ red, P–. Frequent on bark, especially of oaks, statewide, except for the most arid locations, especially drier parts of the Coast Range and the western slope of the northern Sierra Nevada. Until recently, this species was called *P. quercina,* but that species does not occur in North America. Photo from Mendocino Co.

Parmeliopsis

Starburst lichens

Small foliose lichens that form rosettes 1–5 cm wide, with radiating, narrow lobes usually under 1 mm wide. Both species below are sorediate, and apothecia are rare. Resembles *Physcia*, but the spores and chemistry are different, and no *Physcia* has the color of *P. ambigua*. The two species are often found growing together, as in the photo from Malheur National Forest, OR.

Parmeliopsis ambigua

Green starburst lichen

Thallus small, pale, yellow-green, with narrow lobes. Upper surface with granular soredia in irregular, somewhat flattened patches. Lower surface dark brown to almost black. Spot tests: cortex K–, KC+ gold, C–, P–. Occasional, often on wood, sometimes on conifer bark, especially tree bases in sunny locations, rarely on mossy rock, in northern mountains and the Sierra Nevada.

Parmeliopsis hyperopta

Gray starburst lichen

Appearance, range, and habitat much like *P. ambigua*, above, but lobes are bluish gray. Spot tests: cortex K+ yellow, KC–, C–, P+ pale yellow. *Physcia* species that resemble *P. hyperopta* all have a white lower surface.

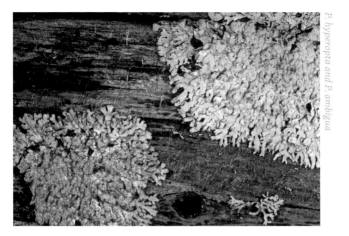

Parmotrema

Ruffle lichens, Scatter-rug lichens

Small to large, broad-lobed, foliose lichens, whitish or pale gray, with ruffled margins that typically have short or long black cilia. Lower surface dark brown or black, especially toward the center, often becoming lighter or blotched white at the margins. Rhizines usually unbranched and almost always growing *only* near the center of the thallus, leaving a broad naked zone close to the margins (but see *P. reticulatum,* below). Some species with large, brown apothecia. In many habitats *Parmotrema* can be common and conspicuous. An additional CA species, *P. cristiferum* (not pictured), has a thallus without maculae but becomes finely cracked in older, central portions; the margins are crinkled, with fine soredia; it lacks cilia, and the underside is black with a broad, brown, bare edge; occasional on the southern coast.

Parmotrema arnoldii

Powdered ruffle lichen

Thallus pale gray, with lobes 6–20 mm wide that have soralia near the margins, making the tips curl downward. Lobes without soredia have long cilia.

Apothecia absent. Lower surface dark brown. Spot tests: cortex K+ yellow; medulla K–, KC+ red to orange, C–, P–, UV+ blue-white. Fairly common on bark, sometimes on rock, mostly in coastal areas, Los Angeles region northward. Compare with *P. hypoleucinum* and *P. perlatum*. Photo from Mt. Tamalpais, Marin Co.

Parmotrema austrosinense
Unwhiskered ruffle lichen
Thallus pale gray, with broad, smooth lobes bearing maculae; margins wavy, edged with fine soredia, without cilia. Lower surface mostly pale to dark brown, with a naked, often white zone at the margin but black in the center. Spot tests: cortex K+ yellow; medulla K–, KC+ red, C+ red, P–. Resembles the more common *P. hypoleucinum* but has smoother lobes without cilia and differs in chemistry. Occasional on bark, rarely on rock, in coastal habitats, Marin Co. to Santa Barbara Co. Photo from central TX.

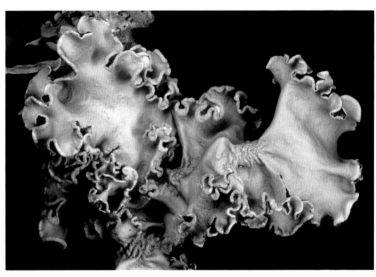

Parmotrema crinitum
Salted ruffle lichen
Thallus pale greenish gray with lobes that are dissected and fringed with cilia. Upper surface with dense, cylindrical or branched isidia that sprout minute cilia at the tips. Apothecia rare. Lower surface dark brown with a bare zone near the margins. Spot tests: cortex K+ yellow turning orange; medulla K+ yellow, KC–, C–, PD+ orange. One of the most strikingly isidiate lichens in CA, found mostly on deciduous trees, sometimes on cedars or other conifers, occasionally on mossy rock, in coastal zones, Monterey Co. northward. Photo from coastal WA.

Parmotrema hypoleucinum
Grainy-edged ruffle lichen

Thallus pale gray to greenish with lobes 3–15 mm wide, often with minute white maculae on the upper surface. Lobes often ascending, their margins typically covered with granular soredia; sparse cilia. Lower surface black and white with a continuous or mottled white bare zone near the margins. Spot tests: upper cortex K+ yellow, KC–, C–, P–; medulla K+ yellow to orange, KC–, C–, P+ orange. On trees in open locations, rarely on rocks, in coastal areas, San Mateo Co. to Mexico. *P. hypotropum* (not pictured), an almost identical species that reacts K+ yellow turning red and P+ deep yellow in the medulla, is rather rare in CA, Marin Co. to Mexico. *P. hypoleucinum* has sparser cilia than *P. arnoldii;* also compare with *P. perlatum*. Photo from the Channel Is.

Parmotrema perlatum

Common powder ruffle

Thallus pale gray to somewhat yellowish with lobes 3–15 mm wide; quite ruffled looking with lobe tips that curl downward slightly; rounded soralia on or near the margins; cilia less than 3 mm long. Lower surface black at the center but brown at the edges. Spot tests: cortex K+ yellow; medulla K+ yellow to orange, KC–, C–, P+ orange. Probably the most common species of *Parmotrema* in CA, found along the entire coast on bark, occasionally on shaded rock. It is smaller and more profusely sorediate than *P. arnoldii;* with shorter cilia, a browner underside, and a smoother lobe surface than *P. hypoleucinum;* it differs in chemistry from *P. stuppeum.* Photo from Santa Barbara Co.

P. perlatum

Parmotrema reticulatum

Thallus fairly large, ruffled-looking, with lobes 4–15 mm wide, a fine network of maculae and cracks, and powdery soralia on or close to the margins. With

P. reticulatum

both cilia and unbranched or squarrose rhizines closer to the margins than on other species of *Parmotrema*. Apothecia rare. Spot tests: cortex K+ yellow, UV–; medulla K+ red, KC–, C–, P+ orange. The conspicuously cracked surface and presence of rhizines almost up to the margins set it apart from otherwise similar species of *Parmotrema*. Uncommon on bark, central and southern coast. Photo from the Channel Is.

Parmotrema stuppeum
Powder-edged ruffle lichen
Thallus pale gray, smooth, without maculae. Lobes broad, 10–15 mm wide, with margins edged with soredia; cilia usually sparse. Lower surface black with a bare brown zone around the edges. Spot tests: cortex K+ yellow, UV–; medulla K+ red, KC–, C–, PD+ yellow to orange. On bark of deciduous trees, occasionally on conifer bark or rock, in coastal zones, Marin Co. to Channel Is. Resembles *P. perlatum* but differs in chemistry. Photo from the Channel Is.

Peltigera
Pelt lichens, Dog lichens
Small to large foliose lichens, most species with broad lobes, upper surface smooth (Dog lichens) or tomentose (Pelt lichens). Lower surface felty, without a cortex, but most species with raised veins and rhizines. Species with *Nostoc* as the photobiont have dark gray or brown thalli; those with the green alga *Coccomyxa* have brownish green to very green thalli; these species

have cephalodia containing *Nostoc,* usually scattered on the upper surface of the lobes. Apothecia on the lobe margins have reddish brown to black disks, often saddle-shaped. Spores fusiform to needle-shaped, colorless, 4- to many-celled. Spot tests usually negative. *Peltigera* species are most often found in moist environments, on soil cut-banks or mossy rock, sometimes on bark, especially around tree bases. The appearance of the lower surface is important in telling the species apart. None of the species of *Peltigera* are truly common in CA, but they can be locally abundant, and the larger ones can be conspicuous, especially when damp.

P. aphthosa

Peltigera aphthosa
Common freckle pelt, Felt lichen

Thallus large-sized, with lobes up to 4 cm wide, bright green in the shade, especially when wet, becoming brownish in the sun. Lobes broad, with scattered grayish brown irregularly shaped, tightly attached cephalodia containing cyanobacteria. Apothecia red-brown, on the margins, with a warty back surface. Lower surface almost uniformly black to dark brown, becoming slightly veined and then white near the margins. Photobiont is a green alga. Occasional on mossy soil, rocks, or tree bases in humid forests in northern CA. Can look very similar to *P. britannica.* Photo from Wells Gray Park, BC.

Peltigera britannica
Flaky freckle pelt

Strongly resembles *P. aphthosa,* except with lobed cephalodia that detach easily. These lobes often break off and develop independently as a separate lichen that is dark brownish gray; this "cyano-lichen" eventually captures some cells of *Coccomyxa* green algae and grows back into a thallus of *P. britannica.* Approximately the same habitat and range in CA as *P. aphthosa* and also uncommon in the state. Both forms can be seen in this photo of a wet lichen from Mt. Hood National Forest, OR; the cyanobacterial form is the dark olive part of the thallus in the center of the picture.

P. britannica

Peltigera canina

Dog lichen

Thallus medium-sized to large, gray or brown, with a fuzzy tomentum, especially near the lobe margins. Lobes 10–25 mm wide, smooth to slightly blistered. Apothecia common on the margins, brown. Lower surface mostly white, becoming brownish near the center, with conspicuous, raised, white veins and pale tufted rhizines. Occasional on soil and moss, in woodlands, open fields, or sandy places, less often on bark or mossy rock, statewide. Resembles *P. rufescens*, but that has narrower, more upturned lobes, heavier tomentum, and shorter rhizines that merge together. Photo from coastal WA.

P. canina

Peltigera collina

Tree pelt

Thallus relatively small, with lobes only 5–10 mm wide, most often dark gray but can be brownish or sometimes lighter gray. Upper surface smooth or a bit scabrose, occasionally pruinose, lobe margins often ruffled and usually with coarse blue-gray soredia. Apothecia uncommon. Lower surface pale with slight veining and slender or tufted rhizines. Most often on tree trunks, sometimes on mossy rocks, occasionally on soil, statewide. One of the more common species of *Peltigera* in CA. Resembles the sorediate *Nephroma parile*, but the combination of veins and rhizines on the lower surface marks it as a species of *Peltigera*. Compare also with *Sticta limbata*, which has round cyphellae on the lower surface. Photo from Malheur National Forest, OR.

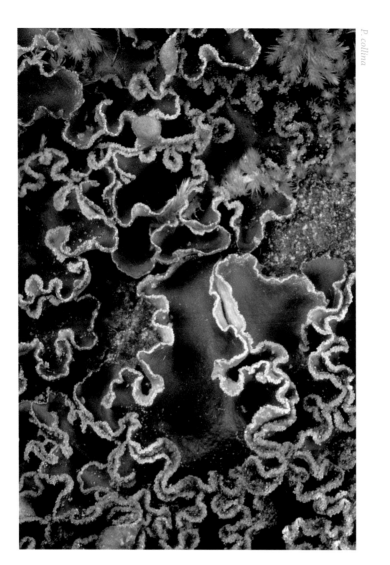

Peltigera didactyla
Alternating dog lichen

Small thallus with rounded, concave lobes up to 1 cm wide that typically have round patches of grayish, granular soredia. As the lichen ages, these patches heal over, leaving a smooth surface, and the thallus develops reddish brown fruiting bodies on the tips of vertically ascending lobes. Lower surface pale cream, with whitish to tan veins and white to brownish rhizines. Occasional on soil, mostly in the Sierra Nevada and northern mountains. Photo from southern BC.

Peltigera gowardii
Waterfan

A small to medium-sized aquatic lichen, dark gray when dry, but purplish brown, and translucent, rather like seaweed, when completely immersed. Apothecia common, small, reddish brown, without margins. The photobiont is *Nostoc*. Rather rare on rocks in mountain streams, central to northern Sierra Nevada. Photo from the Gifford Pinchot National Forest, WA.

Peltigera leucophlebia
Ruffled freckle pelt

Thallus gray to brownish when dry but bright green when wet, with dark gray cephalodia scattered on the surface. Lobes with wavy margins. Apothecia brown with a patchy, discontinuous cortex underneath. Lower surface with

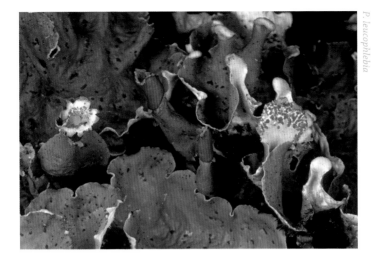

dark brown veins and pale to dark rhizines. A Pacific Northwest species, uncommon on mossy soil, decaying wood, or rock in northwest coastal forests. Photo from Wells Gray Park, BC.

Peltigera membranacea
Membranous dog lichen

Thallus gray to brownish or olive with broad but thin lobes 2–3 cm wide; fairly smooth and sometimes shiny except for the slightly tomentose lobe tips. Apothecia frequent, red-brown, saddle-shaped. Lower surface quite pale with a network of narrow, raised, tomentose veins and long, ropelike, tomentose rhizines. On mossy substrates of all kinds in the Sierra Nevada and Coast Range, statewide. One of the more common *Peltigera* species. The thin lobes with raised, fuzzy tomentum and details of the lower surface set it apart from similar species. Photo from Sonoma Co.

Peltigera ponojensis

Pale-bellied dog lichen

Thallus gray to dark brown, usually heavily tomentose, lobes 5–15 mm wide and upturned at the margins. Apothecia sparse, rounded, and flat. Lower surface pale, with tan to dark brown veins and distinct, pale, unbranched rhizines. Occasional on mossy or bare soil in rather dry sites in the northern Coast Range and Sierra Nevada. Resembles *P. canina,* which has paler veins on the lower surface and more fibrous rhizines, and *P. rufescens,* which is always tomentose and has more elongated apothecia and very short, confluent rhizines. A similar, small species, *P. monticola* (not pictured), has reddish brown veins, tomentum close to the lobe margins, and can become minutely lobulate; occasional in southern mountains. Photo from Humboldt Co.

Peltigera praetextata

Scaly dog lichen, Born-again pelt

Thallus brown to gray, smooth and fairly dull, with a light tomentum near the lobe tips. Lobes 10–30 mm wide; the margins are often finely ruffled, with small lobules and cracks. Apothecia uncommon. Lower surface pale, with light brown veins that darken toward the center and rather long, discrete, ropy to fibrous rhizines. On rock, logs, or soil, statewide, mostly in mountain forests. The lobulate and tomentose lobe margins usually separate this species from the others. Photo of a damp specimen from Olympic National Park, WA.

Peltigera rufescens

Field dog lichen

Thallus gray to brown, most often covered with tomentum, lobes 5–15 mm wide and sharply upturned at the margins. Apothecia frequent, dark reddish brown, saddle-shaped. Lower surface with raised, mostly dark veins

that lighten toward the margins and very short, tufted rhizines that become almost continuous near the center. In open fields and roadsides in the sun, statewide, mostly in inland mountain areas. Compare with *P. canina* and *P. ponojensis*. Photo from San Luis Obispo Co.

Peltigera venosa
Fan lichen

Thallus small, containing green algae. Lobes rounded, green to dark gray-green, 10–15 mm wide. Apothecia conspicuous around the margins, reddish brown to black. Lower surface pale, with raised dark veins radiating from the base of each lobe; granular cephalodia, the exception to the rule of superficial cephalodia in *Peltigera*. Without rhizines, except where the lobe is attached.

A small but distinctive lichen, on soil, often with mosses, in moist, protected spots in northern CA and the Sierra Nevada, sometimes at mid- to fairly high elevations. Photo of a wet specimen from Sonoma Co.

Also of Note

Three other species, not pictured, occur in CA, but they are seldom encountered:

Peltigera degenii is like *P. polydactylon*, below, but with narrow, raised veins; occasional in the northern and central Coast Range.

Peltigera horizontalis is also similar to *P. polydactylon*, but has a grayish thallus with depressions corresponding to the rhizines below and large, round, flat apothecia rather than the saddle-shaped, recurved apothecia of *P. polydactylon*; uncommon in the Sierra Nevada.

Peltigera polydactylon has a shiny, brownish to greenish gray surface, numerous recurved, almost tubular apothecia on upturned lobes, and a lower surface with wide, low veins, creating an almost uniformly brown-black surface with conspicuous large white spots; rare in the Coast Range.

Phaeophyscia

Shadow lichens

Small foliose lichens, thallus brownish or gray to olive, usually with radiating lobes typically less than 1.5 mm wide. Apothecia, if present, are lecanorine with dark brown to black disks, sometimes fringed with black rhizines. Lower surface black (except in *Ph. nigricans*), sometimes with cilia at the margins. Spores dark brown with unevenly thickened walls. Spot tests negative in most species. *Physcia, Physconia,* and *Physciella* (no species pictured) are closely related genera; *Phaeophyscia* species are mostly smaller and darker than the first two and epruinose; *Physciella* has a pale lower surface.

Phaeophyscia ciliata
Smooth shadow lichen

Thallus pale gray to brownish with radiating narrow lobes that have rhizines at the margins resembling cilia. Apothecia common, 0.4–1.5 mm wide, disks almost black with prominent margins. Without soredia or isidia. Occasional on bark, mostly of deciduous trees, sometimes on mossy rock, in mountains statewide, most often inland. Photo from southern MS.

P. ciliata

Phaeophyscia decolor
Starburst shadow lichen

Thallus gray to brownish, pale or dark, sometimes slightly pruinose, with narrow lobes 0.2–0.5 mm wide. Apothecia common, reddish brown to black, less than 1 mm wide. Without soredia or isidia. Fairly common on rock in inland mountains statewide, sometimes in drier Coast Range locations. Compare with *P. sciastra.* Photo from Alpine Co.

P. decolor

Phaeophyscia hirsuta

Bearded shadow lichen

Thallus pale gray to brownish, up to about 4 cm wide, with small lobes 0.5–1 mm wide, sometimes with maculae, and fine colorless hairs on the tips. Soredia fine to coarse, greenish, in soralia that vary from labriform to elongated and may be marginal or laminal. Apothecia rare. Lower surface dark brown to black. On bark and rock, statewide, but most common in southern mountains and coastal areas. Sometimes resembles two much rarer species (not pictured), *Ph. kairamoi,* which has very isidioid soredia, and *Ph. nigricans,* also with granular to isidioid marginal soredia but with a pale lower surface. Photo from Sonoma Co.

P. hirsuta

Phaeophyscia orbicularis

Mealy shadow lichen

Thallus greenish gray to brownish gray with flat to convex lobes. Upper surface with dark greenish gray, powdery to finely granular soredia in irregular soralia, not concentrated on the lobe tips or margins like most other sorediate species. Apothecia uncommon. Medulla white. Usually on bark, sometimes on rock, statewide, in both inland and coastal areas and in the Sierra Nevada. Resembles *P. hispidula* (not pictured), a southwestern species found occasionally in southern CA with concave, upturned lobes and more conspicuous rhizines. Photo from Humboldt Co.

Phaeophyscia sciastra

Dark shadow lichen, Five o'clock shadow

Thallus usually very dark gray, sometimes lighter, with radiating, narrow lobes 0.15–0.5 mm wide, and with coarse, black, granular isidia, or isidia-like soredia, along the margins, especially in the central parts of the thallus. Apothecia rare. Often forms rosettes on rock; specimens without isidia or soredia can resemble *Ph. decolor.* Most common in southern interior mountains but found statewide. Photo from the Rocky Mountains, CO.

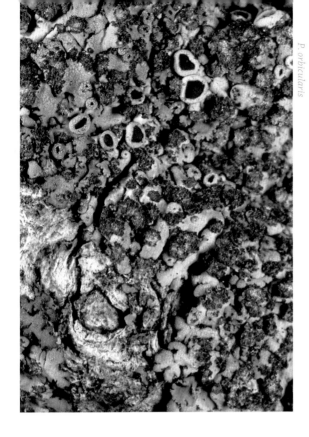

P. orbicularis

P. sciastra

Physcia

Rosette lichens

Medium-sized to small foliose lichens, white to pale greenish gray, often with maculae, but generally epruinose. Lower surface usually pale, rarely brown to black. Rhizines, when present, usually white and near the margins. Spores dark brown, thick-walled with angular or oval locules, 2-celled. Spot tests: upper cortex K+ yellow, KC–, C–, P– or pale yellow; medulla K– or K+ yellow, KC–, C–, P– or pale yellow. *Physcia* species are among the more common lichens in CA. Compare with *Heterodermia, Parmeliopsis, Phaophyscia,* and *Physconia;* the pale lower surface and K+ yellow upper cortex will help separate *Physcia* species from these.

Physcia adscendens

Hooded rosette lichen

Thallus small, pale gray to slightly bluish with narrow, ascending lobes in a cluster that make it look almost fruticose, not forming a flat rosette. Lobe tips with soralia like tiny hoods or helmets, and granular soredia within the hoods; both lobe margins and tips with long cilia, usually white and un-branched. Lobes with white maculae, sometimes pruinose. Apothecia un-common. Spot tests: medulla K–. Fairly common on bark, twigs, and wood, occasionally on rock, statewide, except for the Mojave Desert. Can resemble *Ph. tenella* or *Ph. tenellula,* but those have soralia mainly on the lobe tips, not in hoods. The marginal cilia make it resemble a species of *Heterodermia,* but the hooded soralia are distinctive. Photo from a gravestone in western OR.

P. adscendens

Physcia aipolia

Hoary rosette lichen

Foliose thallus up to 5 cm wide, with narrow, radiating, sometimes overlap-ping lobes, pale to dark gray, spotted with white maculae but without soredia or isidia. Apothecia common, 1–3 mm wide, dark brown but usually covered with white pruina. Lower surface white to tan with numerous pale rhizines. Spot tests: cortex and medulla K+ yellow. Quite common on bark and wood in open habitats statewide, both coast and inland. Resembles *Ph. stellaris,* but that species lacks maculae and reacts K– in the medulla; also resembles *Ph. phaea* except for the pruinose apothecia. Could perhaps be confused with *Parmelina coleae,* but that species has a black lower surface, little or no pruina, and is K– in the medulla. Photo of a damp specimen from Monterey Co.

P. aipolia

Physcia albinea
Small-spotted rosette lichen

The name *Physcia albinea* is problematic and is used here "in the sense of North American authors," referring approximately to a non-sorediate rock lichen related to the sorediate *Ph. tribacia* (both have a lower cortex made up

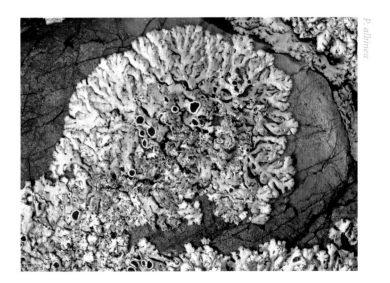

P. albinea

of pseudoparenchyma with round cells), which used to be called *Ph. callosa*. *P. albinea,* as the name is used here, has a white to pale gray thallus forming rosettes with downturned lobes 0.4–2 mm wide that become lobulate at their tips, especially near the center of the thallus. Apothecia small, dark brown to black; without soredia or isidia. Spot tests: medulla K–. It has much broader, more convex lobes than the similar eastern species, *Ph. halei.* It may be fairly common and widespread, but until the species of *Physcia* in western North America are more fully understood, its true range will remain unknown. Photo from coastal rocks, Marin Co.

Physcia biziana
Frosted rosette lichen
Thallus pale to dark gray, dull, coarsely pruinose, up to 3 cm wide with lobes of variable width, 1–5 mm, that are only loosely attached and often overlapping. Lobe tips can be flat or convex and often turn down. Maculae absent. Apothecia common, dark brown, but usually covered with a coarse, white pruina. Lower surface white to tan with scattered white to gray rhizines. Spot tests: upper cortex K+ yellow, P+ yellow; medulla K– or faintly yellowish. On bark in exposed habitats statewide. One of the most common *Physcia* species in central and southern mountain areas. Specimens with only thin pruina might be mistaken for *Ph. stellaris,* which has narrower lobes. Compare with its sorediate counterpart, *Ph. dimidiata.* Photo from the Mojave Desert.

P. biziana

Physcia caesia
Blue-gray rosette lichen, Powder-back rosette
Thallus pale gray, up to 5 cm wide, darkening in the center, with convex lobes 0.5–3 mm wide spotted with maculae, sometimes faintly; conspicuous rounded clusters of bluish gray granular soredia dot the upper surface. Apothecia rare. Lower surface tan to brown with short, dark rhizines. Spot tests: upper cortex and medulla K+ yellow, P+ yellow. This is the sorediate

counterpart of *Ph. phaea* and is quite common. A similar but rare species in central to southern coastal areas, *Ph. poncinsii,* has lobes with truncate tips. On rock and wood in exposed habitats, rarely on bark, statewide, especially inland. Photo from Lassen National Forest.

P. caesia

Physcia dimidiata
Halved rosette lichen

Similar in most respects to *P. biziana,* but with granular soredia that are mostly marginal (but can also occur on the surface of outer lobes). Apothecia absent. Spot tests: upper cortex K+ yellow, P+ yellow; medulla no reactions. Fairly common on rock, sometimes on wood and bark, mostly coastal ranges from San Francisco to Mexico, but also in the Sierra Nevada and southern interior. Photo from the Mojave Desert.

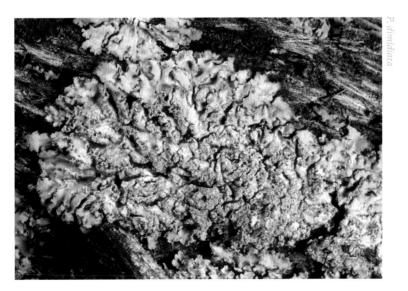

P. dimidiata

Physcia dubia
Powder-tipped rosette lichen

Thallus small, up to 3 cm wide, pale to dark gray or sometimes yellowish, with convex lobes that fan out at the margins; the tips curl up, with lip-shaped soralia on the underside bearing gray to greenish soredia. Lobes sometimes have white maculae or a light coating of pruina. Apothecia rare. Spot tests: upper cortex K+ yellow, P+ yellow; medulla no reactions. Fairly common on rock or wood, occasionally on bark, statewide, especially in the Sierra Nevada and inland ranges. Compare with *Ph. tribacia,* which is typically larger, shiny, and with downturned lobe tips. Somewhat resembles *Ph. millegrana,* very rare in coastal CA, which has finely divided lobe tips producing coarsely granular soredia on the edges and with frequent apothecia. Photo from the Mojave Desert.

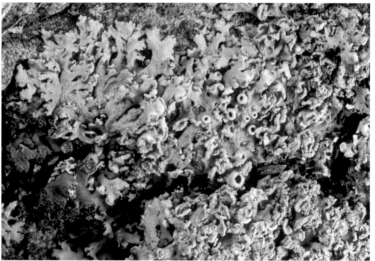

Physcia erumpens
Bursting rosette lichen

Thallus up to 3 cm wide and usually adnate, pale gray to bluish or brownish with lobes up to 1 mm wide that are separate or overlapping, occasionally pruinose. Upper surface with soralia-like small craters filled with greenish soredia. Lower surface black with lighter tips and black rhizines. Spot tests: upper cortex and medulla K+ yellow, P+ yellow. On rock, uncommon, central CA coastal areas. Photo from Marin Co.

Physcia phaea
Black-eyed rosette lichen

Thallus pale to dark gray or brownish, up to 3 cm wide with crowded, often overlapping, convex lobes up to 1.5 mm wide and conspicuous maculae, often rugose. Apothecia usually numerous, round, slightly raised, black, and epruinose; thallus without lobules, pruina, soredia, or isidia. Lower surface

tan to brown with brownish rhizines. Spot tests: upper cortex and medulla K+ yellow, P+ yellow. Fairly common on rock, especially granite, statewide. *Ph. phaea* is the fertile counterpart of *Ph. caesia;* it also resembles *Ph. aipolia,* which grows on bark and typically has pruinose apothecia. Photo from the Channel Is.

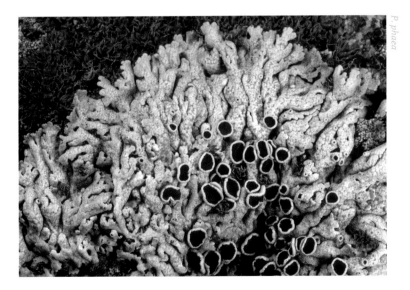

Physcia stellaris
Star rosette lichen

Thallus pale to dark gray, up to 4 cm wide, mostly uniform but occasionally with some spotting. Lobes radiating, 0.5–1.5 mm wide, flat in the center to convex at the tips, sometimes crowded. Apothecia common, dark brown but often with pruina; without soredia or isidia. Lower surface white to tan with pale or dark rhizines. Spot tests: upper cortex K+ yellow, P+ yellow; medulla no reactions. Resembles *Ph. aipolia,* but that has many conspicuous maculae, lobes that are flat to slightly concave at the tips, and a medulla that is K+ yellow. Fairly common on bark, especially of broad-leaved trees, rarely on wood or rock, statewide, especially in the Coast Range, notably on shrub oaks in the southern part of the state. Photo of a damp specimen from southern ON.

P. stellaris

Physcia tenella
Fringed rosette lichen
Thallus small, with narrow, ascending, branched, often tangled lobes 0.2–1 mm wide, white to pale gray or sometimes yellowish or darker, usually

P. tenella

uniform but occasionally spotted with maculae. Lobes with upturned tips and lip-shaped soralia, fringed with both marginal cilia and rhizines. Apothecia fairly common, dark, red-brown, sometimes pruinose, about 1 mm wide. Spot tests: medulla K–. On twigs, bark, and rock, statewide, both coast and inland, except Mojave Desert. Somewhat resembles *Ph. adscendens,* but that species has hooded soralia and infrequent apothecia; also resembles *Ph. tenellula* (not pictured), a smaller species with white to black hairs on the upper surface found in southern CA coastal areas. Photo from Plumas National Forest.

Physcia tribacia
Beaded rosette lichen
Thallus round to irregular, up to 3 cm wide, with shiny, narrow, white to gray, downturned lobes up to 2 mm wide. Apothecia occasional, dark brown or black, constricted at the base and with partly sorediate margins. Lower surface with abundant soredia on the lobe tips, often eroding them; without maculae. Spot tests: upper cortex and medulla K+ yellow, P+ yellow; medulla no reactions. Fairly common on exposed rock, especially calcareous or where covered with bird droppings, statewide, especially in coastal habitats, central and southern CA. North American material of *Ph. tribacia* has long been called *Ph. callosa.* Photo from the Channel Is.

P. tribacia

Also of Note
An additional CA species is *Physcia clementei* (not pictured), which has an adnate thallus that is fragile, gray, often sorediate in the center and apothecia with crenulate margins; fairly rare, on bark and rock along the central and southern coast.

A related genus, not pictured, is *Physciella;* one species, *Ph. chloantha,* is rare in CA. It has a small, gray to pale brown, sorediate thallus with a K– cortex, rather like a species of *Phaeophyscia* but with a pale lower surface.

Physconia
Frost lichens

Small to medium-sized foliose lichens, pale gray to greenish or brown, usually turning green when wet; always frosted with pruina, at least on the lobe tips. Apothecia dark brown, lecanorine, usually pruinose; many species with soredia. Lower surface black or, occasionally, pale, with abundant, branched, squarrose rhizines (rarely unbranched or forked) that often create a fringe around the thallus. Medulla white to pale yellow. Spot tests: upper cortex negative; medulla and soredia reactions vary by species. Species of *Physconia* can resemble *Phaeophyscia*, except for the pruina and squarrose rhizines, and *Physconia* species are generally larger.

Physconia americana
Fancy frost lichen

Thallus pale grayish brown, distinctly green when wet, up to 10 cm wide, with long, narrow lobes 1–2 mm wide that usually radiate and sometimes overlap. Upper surface pruinose, especially on the lobe tips. Apothecia almost always present. Lobules frequent around the apothecial margins and/or on older parts of the thallus; without soredia or isidia. Lower surface tan to black. Spot tests negative. Quite common, usually on bark, sometimes on rock, statewide, except for the Mojave Desert. Somewhat resembles the less common *Ph. muscigena*, but that species tends to be darker, with upturned peripheral lobes, and grows on mossy soil. Also resembles the much rarer *Ph. californica*, which is more lobulate and has a paler lower surface. Photo of a damp specimen from Lassen National Forest.

P. americana

Physconia californica
California frost lichen

Thallus gray to brownish, up to 7 cm wide, without soredia or isidia, but with abundant lobules. Completely or partially pruinose, sometimes with apothecia. Lower surface pale to tan with rhizines. Spot tests negative. Uncommon on bark in coastal and inland areas, central to southern CA. Photo from the Mojave Desert.

Physconia enteroxantha
Yellow-edged frost lichen

Thallus small, up to 6 cm wide, with radiating lobes 0.6–3 mm wide, pale gray-brown to brown or greenish, and with conspicuous white pruina, sometimes only on the lobe tips. Lobe margins edged with yellowish green, granular soredia, especially pronounced near the central, older parts of the thallus. Apothecia rare. Lower surface pale at the margins but dark at the center with black, squarrose rhizines. Medulla usually pale yellow, rarely white. Spot tests: medulla K+ yellowish but sometimes hard to see, KC+ yellow to orange, C–, P–; soralia usually the same but sometimes K–, KC–. On bark and wood, occasionally on mossy rock. *Ph. enteroxantha* is the most common species of the genus in CA, especially abundant on trees in drier habitats. In cases where the medulla is white, it might be mistaken for other sorediate species, such as *Ph. isidiigera*. Resembles *Ph. fallax* (not pictured), which has ear-shaped soralia and a K– medulla; it occurs in central and southern coastal regions and in the Sierra Nevada. Many older records of *Ph. detersa* (which probably does not occur in CA) are likely to be *Ph. enteroxantha*. Photo from San Bernardino National Forest.

Physconia isidiigera

Grainy frost lichen

Thallus gray to brown, pale or dark, up to 7 cm wide, lobes 0.6–3 mm wide, flat or slightly upturned, somewhat dissected, pruinose all over or only on the tips. Marginal soralia have granular soredia that are sometimes corticate and elongated, resembling isidia, especially abundant near the center of the thallus. Apothecia occasional. Lower surface tan to black, especially in the central area, with squarrose rhizines. Spot tests negative. Similar to *Ph. enteroxantha*,

which usually has a yellowish medulla and differs in chemistry. Fairly common on bark and wood, occasional on rock, statewide. Photo from San Bernardino National Forest.

Physconia muscigena
Ground frost lichen

Thallus small to medium-sized, with dissected, overlapping, and ascending concave lobes, pale greenish to purplish brown, often varying on the same thallus. Pruinose, especially toward the center; without soredia or isidia. Apothecia, if present, with smooth or lobulate margins. Lower surface black with closely packed, black, squarrose rhizines. Spot tests negative. Occasional on the ground over moss, plant debris, or rock, usually in mountain areas, especially inland. Photo from southern BC.

P. muscigena

Physconia perisidiosa
Crescent frost lichen

Thallus gray to brownish or dark brown, up to 5 cm wide, with fairly short, overlapping lobes 0.5–1.5 mm wide. Usually with heavy pruina; edges of the lobes have crescent-shaped soralia with coarse soredia that can resemble isidia. Center of the thallus sometimes with small convex lobules. Apothecia rare. Lower surface black or pale at the tips, fibrous, lacking a cortex, with squarrose rhizines. Spot tests negative. The marginal, lip-shaped soralia give it a different appearance from *Ph. enteroxantha* or *Ph. isidiigera*, which typically have soredia produced all along the margins. Common on bark, sometimes on soil or mossy rock, statewide, mostly inland but can be coastal. Resembles *Ph. leucoleiptes* (not pictured), an uncommon species of the central and southern mountains with a smooth, corticate lower surface, white at the margins then darkening, and crescent-shaped soralia that react K+ yellow, KC+ yellow to orange. Photo from San Bernardino National Forest.

P. perisidiosa

Platismatia
Rag lichens
Medium to large foliose lichens with crinkled, ascending lobes (some species quite shrubby), pale greenish white to slightly brown. Often with isidia, less commonly with soredia or apothecia. Lower surface brown to black, sometimes with white patches near the margins; may have sparse, usually unbranched rhizines. Spores colorless, 1-celled, 8 per ascus. Spot tests: cortex K+ yellow, KC–, C–, P–; medulla no reactions. Mostly on conifer bark in forests.

Platismatia glauca
Varied rag lichen, Ragbag
Thallus foliose with rather broad, suberect, irregular lobes that can be quite variable in width (3–20 mm), greenish white, sometimes browning at the edges, with margins often finely divided, bearing isidia, soredia, or a mixture of both. Lobes sometimes with branched, subfruticose, often isidiate outgrowths. Apothecia and pycnidia very rare. Lower surface black in the center, mottled brown and white at the edges. Fairly common on bark in Douglas-fir forests, occasionally on wood, rock, or soil, northern Coast Range and interior mountains and western slope of the Sierra Nevada. Compare with *Esslingeriana idahoensis*. Photo from Olympic National Park, WA. A similar species, *Platismatia wheeleri* (not pictured), lacks the isidioid projections from the margin that often develop in *P. glauca* and has long, lineal, marginal, powdery soralia. *P. wheeleri* grows in moist areas, such as riparian corridors, as far south as Palomar Mountain.

Platismatia herrei

Tattered rag lichen

Thallus pale greenish or bluish gray, with long, narrow lobes rarely more than 3 mm wide, flat or inwardly curled. Lobe margins finely divided, with abundant isidia that can also grow on the lobe surface. Lower surface black to brown, with white patches at the tips, rarely with rhizines. On branches and twigs of conifers, mostly coastal forests, central and northern CA. Photo from the eastern side of the Cascades, OR.

Platismatia stenophylla

Ribbon rag lichen

Thallus pale gray to greenish gray, greener when wet. Lobes branched, narrow, 0.5–4 mm wide, in clumps, dividing near the tips and often brown at the edges; sometimes spotted with maculae; without soredia or isidia. Apothecia occasional, red-brown, at the lobe tips; usually with black pycnidia on the lobe margins. Lower surface black at the center, becoming brown and then white at the tips, without rhizines. On bark, mostly of conifers, in northern coastal forests. Photo from near Willits, Mendocino Co.

P. stenophylla

Pseudocyphellaria

Specklebelly lichens

Medium to large foliose lichens with broad, loosely attached lobes that have a network of ridges and depressions. Lower surface tan, tomentose, with conspicuous, often raised, white or yellow pseudocyphellae. Some species frequently have apothecia, others soredia; some contain cyanobacteria (including those described below), others have green algae. Medulla white or yellow. Spores colorless, 2- to many-celled with pointed ends, 8 per ascus. Spot tests negative. On bark, most often of deciduous trees, occasionally on mossy rock, in Coast Range forests.

Pseudocyphellaria anomala

Thallus medium to large, up to 20 cm wide, with broad, reddish brown to gray-brown lobes and lines of pale gray to white soredia on the ridges, sometimes between the ridges as well. Can become almost black when wet. Apothecia uncommon. On bark, especially of broad-leaved trees, such as bigleaf maples. The species is the sorediate counterpart of *P. anthraspis*. Somewhat resembles *Lobaria pulmonaria,* which is typically somewhat greener (primary photobiont is green algae) and lacks pseudocyphellae on the lower surface. Rather common, most frequently in the central and northern Coast Range, occasionally in humid habitats in northern interior mountains as far south as Santa Barbara Co. Photo from Sonoma Co.

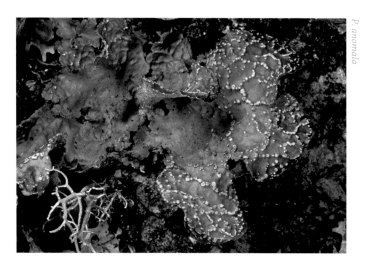

P. anomala

Pseudocyphellaria anthraspis

Thallus large, with broad, reddish brown to gray-brown lobes, becoming dark brown when wet. Apothecia abundant on upper surface, reddish brown, turning bright orange when wet. The fertile counterpart of *P. anomala,* with a very similar habitat and range, south to Los Angeles Co. Photo from Six Rivers National Forest.

P. anthraspis

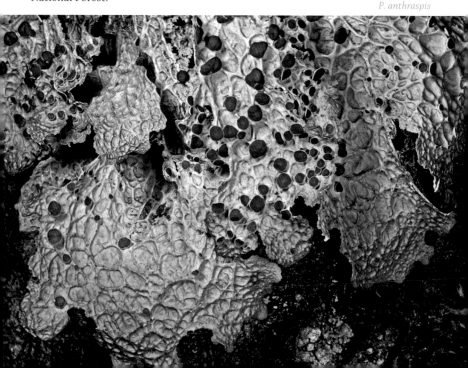

Pseudocyphellaria crocata

Thallus medium-sized, dark shiny brown to brownish gray, with conspicuous bright yellow soredia on the lobe margins and interior surface. Photobiont is a cyanobacterium. Lower surface tomentose with yellow pseudocyphellae. Medulla mostly white, but with some yellow parts near the soralia and pseudocyphellae. The bright yellow soredia make this species unmistakable. Rare on bark of trees of all kinds and on shrubs such as manzanita; occasional in northern coastal forests. The closely related species, *P. perpetua* (not pictured), has strictly marginal soralia and a completely yellow medulla, but its known distribution begins just north of California. Photo from coastal OR.

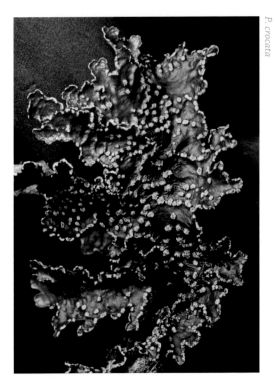

P. crocata

Punctelia

Speckled shield lichens

Medium-sized foliose lichens, pale gray or slightly bluish or brownish, with lobes 3–10 mm wide, dotted with white pseudocyphellae on the upper surface. Apothecia lecanorine but rare in CA species. Medulla white. Spores ellipsoid, colorless, 1-celled, 8 per ascus. Resembles *Flavopunctelia*, but that genus is distinctly green.

Punctelia jeckeri

Powdered speckled shield lichen, Forest speckleback

Thallus pale bluish gray, turning greener when wet, often with a bit of brown at the edges, lobes 2–6 mm wide, becoming thicker and wrinkled toward the center of the thallus. Pseudocyphellae rather inconspicuous, but with soralia containing white to pale gray soredia along the lobe margins and surface, especially in the central parts of the thallus. Apothecia rare. Lower surface tan to dark brown with a mat of short, slender rhizines. Spot tests: K–, KC+ red, C+ red, P–. Common on bark, occasional on rock, in the central and northern Coast Range and oak woodlands. Often found in the same habitats as *Flavopunctelia flaventior* and *Flavoparmelia caperata*. Photo from Mt. Graham, AZ.

P. jeckeri

Punctelia stictica

Rock speckled shield lichen, Seaside speckleback

Thallus closely attached, varying from pale greenish or bluish gray to brownish gray or brown, with lobes mostly 1.5–3 mm wide, bearing conspicuous, raised, irregular, white pseudocyphellae, usually with brown rims that sometimes turn into laminal soralia with coarse, granular soredia; epruinose. Apothecia rare. Lower surface black, except brown in a narrow band at the

P. stictica

margins. Spot tests: K–, KC+ red, C+ pink, P–. Occasional on exposed rock, statewide, Coast Range. Photo from Vancouver Is., BC.

Also of Note
An additional CA species, *Punctelia borreri* (not pictured), has a bluish gray thallus that has conspicuous white pseudocyphellae and granular, pale gray soredia that originate from the pseudocyphellae; spot tests: upper cortex K+ yellow, C–; uncommon on bark in coastal areas, San Luis Obispo Co. into southern CA.

Rhizoplaca
Rock-posy lichens, Rockbright
Small to medium-sized, mostly umbilicate (attached to the substrate with a central holdfast) lichens and therefore mostly foliose, but often giving a crustose appearance; pale yellowish green to tan or pale gray, with thalli that are usually compact and lumpy, often mounding above the substrate; with lecanorine apothecia. Spores ellipsoid, small, colorless, 1-celled. On rock, sometimes soil, in desert and Great Basin habitats.

Rhizoplaca chrysoleuca
Orange rock-posy
Thallus pale yellowish green or yellowish gray, usually thick and lumpy with lobules and warts. Apothecia orange to somewhat tan, rimmed, lightly pruinose. Spot tests: upper cortex K– or yellow, KC+ yellow-orange, C–, P–; medulla K–, KC– or red, C– or red, P– or yellow. The lumpy, umbilicate thallus and color of the apothecia make this species easy to identify. Fairly common on rock in arid habitats statewide, inland mountains and deserts, almost never coastal. Photo from Arapaho National Forest, CO.

R. chrysoleuca

Rhizoplaca marginalis
White rock-posy
Thallus small, 0.7–1.5 cm wide, smooth or wrinkled, grayish yellow but appearing uniformly white from pruina. Apothecia dark brown to black but

often look gray from pruina along or near the margins. Often with numerous pycnidia. Spores almost spherical to broadly ellipsoid, 9–14 x 6.5–8.5 μm. Spot tests: upper cortex K+ pale yellow, KC+ yellowish, C+ bright yellow, P–; medulla no reactions. A rare lichen of southern CA desert regions. A similar, but smaller and somewhat more common species, *R. glaucophana* (not pictured), has fewer pycnidia, much longer and narrower, oblong, somewhat curved spores, and reacts K–; it shares roughly the same habitat. Photo from the foothills of the southern Sierra Nevada, Kern Co.

R. marginalis

Rhizoplaca melanophthalma
Green rock-posy

Thallus pale greenish or yellowish green, similar in structure to *R. chrysoleuca*, but variable, with apothecia that are usually dark green, sometimes almost black but occasionally more tan, and pruinose. Spot tests: variable, usually

R. melanophthalma

cortex KC+ yellow; medulla K–, KC–, C– or occasionally red, P+ yellow, some-times P–. Common on rock, in about the same habitats and with the same range as *R. chrysoleuca*. More appressed thalli can look almost crustose and can resemble some species of *Lecanora,* especially *L. novomexicana,* or even *L. muralis,* but *Lecanora* species are not umbilicate. Photo from Inyo Co.

Rhizoplaca peltata
Brown rock-posy
Thallus thick and stiff, pale yellow-green, broken into areoles with black edges and numerous cracks. Apothecia usually abundant, most often tan but can be greenish, especially if damp; epruinose. Lower surface pale to dark brown at the center, becoming almost black at the margins, puffy, and with a network of white cracks. Spot tests: upper cortex KC+ yellow; medulla K–, KC–, C–, P+ orange. On rock, uncommon in CA, mostly in southern deserts. Photo from UT.

R. peltata

Rhizoplaca subdiscrepans
Scattered rock-posy
Thallus of scattered thick areoles, dull yellowish green or somewhat gray. Apothecia abundant, orange-tan. In color it resembles *R. chrysoleuca.* Spot tests: cortex KC+ yellow; medulla usually no reactions, rarely K+ yellow, C+ red, or P+ yellow. On rock in arid localities, occasional on the eastern side of the Sierra Nevada and southern CA mountains. Like *R. peltata* or *R. melanoph-thalma,* it can sometimes resemble species of *Lecanora.* Photo from Joshua Tree National Park.

Sticta
Moon lichens, Crater lichens
Medium-sized foliose lichens, dark in color, some species with soredia or isidia. Apothecia very rare. Lower surface corticate, tomentose, with round,

white cyphellae like small craters. Photobiont is *Nostoc*, a cyanobacterium. Spot tests negative. On bark, especially of broadleaved trees, or mossy rock, in moist forest habitats.

Sticta fuliginosa
Peppered moon lichen

Thallus very dark brown or dark gray to black or greenish with lobes bearing cylindrical to coralloid isidia that give the surface a granular look. Lobes usually rounded but can sometimes be rather ragged. Apothecia rare. Lower surface tan, with a white tomentum. Medulla white. On mossy bark, especially of deciduous trees such as bigleaf maples, or on mossy rock, in the Coast Range, statewide, more common toward the north. Photo from the Gifford Pinchot National Forest, WA.

Sticta limbata
Powdered moon lichen

Brown to grayish rounded and somewhat ruffled lobes, usually shiny, with pale granular soredia on and near the margins. Lower surface yellowish tan, with grayish to tan tomentum and short, brushlike rhizines. On mossy bark and rock in coastal forests, but it is a northern species, not common in CA. Compare with *Peltigera collina* (veins on the underside) and *Nephroma parile* (lower surface smooth and tan). Photo from the Gifford Pinchot National Forest, WA.

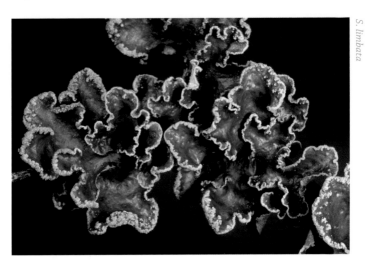

S. limbata

Tuckermannopsis
Wrinkle-lichens, Ruffle lichens

Small to medium-sized foliose lichens with ascending lobes, brown to olive or sometimes pale green, often wrinkled on the lobe surface and ruffled on the margins. Apothecia lecanorine. Lower surface usually light brown with scattered, unbranched rhizines. Frequently with black pycnidia on the lobe margins. Spores simple, spherical, colorless, 8 per ascus. Species can sometimes resemble those of *Melanelia, Melanelixia,* and *Melanohalea,* but those have lobes that are usually closely appressed, not ascending. Mostly on bark and twigs, especially of conifers. One species not described here, *T. sepincola,* is rather small, with smooth brown lobes and numerous apothecia; fairly rare in northeastern CA.

Tuckermannopsis chlorophylla
Powdered wrinkle-lichen

Thallus deep brown, sometimes greenish, with rather narrow, divided lobes 0.7–2.5 mm wide with whitish soredia along the margins and sometimes on the lobe surface as well. Apothecia rare. Lower surface tan, with sparse rhizines. Spot tests negative. Can resemble *Nephroma parile,* which is more appressed, with coarser soredia and no rhizines. Scattered on twigs and branches in conifer forests statewide. Photo from Shasta-Trinity National Forest.

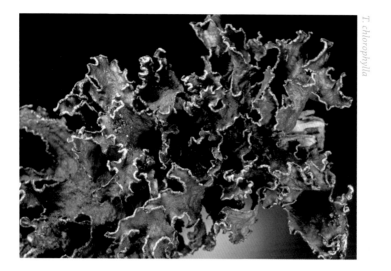

T. chlorophylla

Tuckermannopsis orbata

Variable wrinkle-lichen

Thallus usually somewhat greenish, from pale to dark olive or more brownish, paler in shaded sites, with ruffled lobes 1.5–3.5 mm wide. Margins have black pycnidia, branched lobules, or dark cilia, or any combination of these; when lobules are branched and have pycnidia, they can resemble isidia. Apothecia common, and may originate on the upper surface or the upper or lower sides of lobe margins. Spot tests negative. The most variable species in the genus. On bark, usually of conifers, in coastal and inland forests, statewide, except for the Mojave Desert. Photo from Armstrong Redwoods State Park, Sonoma Co.

T. orbata

Tuckermannopsis platyphylla
Broad wrinkle-lichen

Thallus with rather broad lobes, 3–10 mm wide, usually dark reddish brown becoming greenish when wet, often very wrinkled and with tubercules and lobules that make the lichen look almost isidiate; the tubercules can have white pseudocyphellae. Apothecia frequent and large. Spot tests: medulla no reactions in white areas but K+ orange or yellow in spots and KC+ orange where medulla is yellow, such as in the apothecial margins. On conifer twigs, in northern to central CA forests, usually in drier environments. Very dark specimens might be mistaken for *Kaernefeltia merrillii*. Photo from Plumas National Forest.

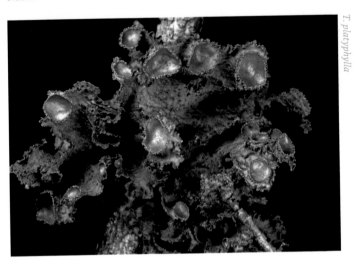

Umbilicaria
Rock tripe

Small to medium-sized foliose lichens, always on rock, sometimes forming patches that cover a large area. Thalli usually gray or brown, pale to almost black, attached by a central holdfast that often can be seen as a bump (umbo) on the surface. Lower surface smooth or wrinkled, sometimes with rhizines. Apothecia lecideine; in some species, ridged and wrinkled. Spot tests: medulla usually K–, KC+ red, C+ pink, P–. Widespread and common on siliceous rock, especially in mountain areas. Can resemble *Dermatocarpon*, but that genus has perithecia instead of apothecia.

Umbilicaria americana
Frosted rock tripe

Thallus 2–12 cm wide, pale gray, or brownish but with coarse, white pruina. Lower surface with a mat of black rhizines coated with black granules. Apothecia uncommon. Occasional in the central and southern Sierra Nevada and southern CA mountains. This is the lightest-colored species of *Umbilicaria* in CA, except for *U. vellea* (not pictured), which is somewhat smaller and distinguished mainly by having branched rhizines that are mostly whitish and naked, interspersed with shorter black rhizines bearing black granules. *U. vellea* is usually found at higher elevations than *U. americana.* A much rarer species, similar to *U. vellea,* with a gray upper surface, is *U. cinereorufescens* (not pictured), with short, thick, black rhizines; Coast Range and Sierra Nevada. Photo from the Salmon River Canyon, ID.

U. americana

Umbilicaria hyperborea
Blistered rock tripe

Thallus brown to very dark, with even or deeply lobed margins, upper surface smooth to very bumpy or verrucose, sometimes with smooth patches between. Apothecia black, with complex patterns of ridges. Lower surface tan to almost black, mostly smooth and usually without rhizines. Fairly common in mountains, especially inland ranges, statewide. Sharing the same habitats,

Umbilicaria hyperborea. Photo from Southeast AK.

The ridged apothecia of *Umbilicaria hyperborea*.

but rare, is *U. proboscidea* (not pictured); it has an upper surface with radiating ridges and apothecia with concentric ridges.

Umbilicaria krascheninnikovii
Salty rock tripe

Thallus dark grayish brown to black, 1.5–4 cm wide, stiff and thick, with conspicuous ridges on the surface that are encrusted with white deposits of dead cells resembling salt. Apothecia black, appressed or slightly stalked, with striking patterns of complex ridges. Lower surface smooth or slightly roughened, tan to pinkish gray, pruinose at the margins, usually without rhizines. Fairly common in mountains, especially inland ranges, statewide. Resembles *U. virginis,* but that species has long rhizines. Photo from Pine Mountain, Ventura Co.

U. krascheninnikovii

Umbilicaria phaea

Emery rock tripe

Thallus usually tan or brown and fairly smooth, with broad lobes up to 6 cm wide. Apothecia black, sometimes circular but often angular or star-shaped, and partially embedded with distinctive ridges forming concentric patterns. Lower surface rough, papillate, pale, rarely with rhizines. The most common species in CA; coast and inland, statewide, tolerant of very arid environments. The rare, brilliant red variety *coccinea* occurs in a few sites in northern CA. A similar, much less common species, *U. angulata* (not pictured), has white, slender rhizines; it occurs statewide in interior mountains.

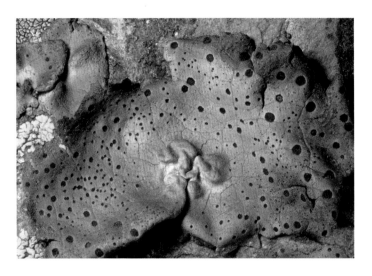

A typical thallus of *Umbilicaria phaea*. Photo from near Lake Isabella, southern Sierra Nevada foothills.

Umbilicaria phaea var. *coccinea*. Photo from near Yreka, Siskiyou Co.

Umbilicaria polyphylla

Petaled rock tripe

Thallus, very dark, 2–6 cm wide, thin and brittle, irregular in outline with deeply incised, smooth, crowded, ascending lobes that sometimes overlap. Apothecia rare. Lower surface sooty black, smooth or with minute papillae, without rhizines. Spot tests: KC+ red, C+ red. Occasional in the Sierra Nevada and northern interior ranges. Photo from near Tok, interior AK.

U. polyphylla

Umbilicaria torrefacta

Punctured rock tripe

Thallus brown to almost black, 2–6 cm wide, often thin and shiny, usually with thin, broken areas and a tattered margin frequently perforated with irregular holes. Upper surface with somewhat convex areoles separated by smooth black areas. Apothecia black, with concentric ridges on the disk.

U. torrefacta

Lower surface tan, roughly papillate, with a network of plates that can become fringed. Spot tests: medulla K+ yellow, KC– or pink, C– or red, P+ orange. In the Sierra Nevada and interior mountains statewide. Photo from BC.

Umbilicaria virginis
Blushing rock tripe

Thallus gray to brownish or very dark, pruinose or with white dead cells that look like salt crystals, especially near the center. Usually very ridged in the middle, becoming more uniform at the margins; Apothecia common, black, often with a ring or button in the center of the disk. Lower surface tan or pinkish, but often brown at the margins, smooth to slightly rough, and with long, slender, sparsely branched rhizines, mostly around the edge. The upper surface can resemble *U. krascheninnikovii,* but that typically lacks rhizines. Statewide, mostly in interior ranges. Photo from Inyo Co., alpine Sierra Nevada.

U. virginis

Also of Note
An additional CA species is *Umbilicaria polyrhiza* (not pictured), which is brownish above, with a lower surface that has plates of tissue around the attachment point; apothecia rare; occasional in Coast Range and Sierra Nevada.

Vulpicida canadensis
Brown-eyed sunshine lichen

Medium-sized foliose lichen with brilliant yellow, ascending, wrinkled lobes. Apothecia numerous, brown, lecanorine, up to 7 mm diameter; pycnidia black, immersed, appearing as black dots; without soredia. Lower surface yellow but paler, often ridged and wrinkled; few or no rhizines. Medulla bright yellow. Spores ellipsoid, colorless, 1-celled, 8 per ascus. Spot tests: cortex K–, KC+ yellowish, C–, P–; medulla K+ yellowish, KC–, C–, P–. Common in drier forests, especially in pine forests on the eastern side of the Sierra Nevada and in northeastern ranges. In color can resemble *Letharia columbiana,* which is fruticose and shrubby with a white medulla. Photo from Wenatchee National Forest, WA.

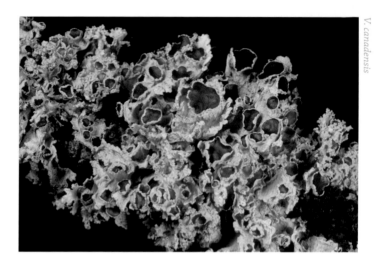

V. canadensis

Xanthomendoza
Sunburst lichens, Orange lichens
Small, bright orange, foliose lichens, often sorediate, sometimes with lecanorine apothecia. Spores polarilocular, 8 per ascus. Spot tests: upper cortex K+ purple, KC–, C–, P–. On bark, wood, and rock, especially if calcareous. Until recently, species in the genus were included within *Xanthoria,* but *Xanthomendoza* species have rhizines (or scattered holdfasts) and a different type of conidia. Some species are often seen in nitrogen-rich environments, such as places where birds perch or agricultural areas where dust from fertilizer is present.

Xanthomendoza fallax
Hooded sunburst lichen
Thallus small, yellow-orange to red-orange, up to 3 cm wide, with appressed or raised lobes that are often somewhat divided at the tips. The tips form crescent-shaped or hood-shaped "bird nest" soralia in a split between the upper and lower cortex, with greenish yellow powdery soredia. Apothecia occasional, up to 1 mm wide; pycnidia uncommon, immersed. Lower surface white, with white rhizines. Fairly common on bark, especially oak, rarely on wood or rock, statewide. Photo from Glenn Co.

Xanthomendoza fulva
Bare-bottomed sunburst lichen
Thallus small, dark reddish to medium orange, up to 9 mm wide, with rounded or divided lobes that form mealy soredia on the lower surface; without hoods.

Thalli sometimes coalesce into larger colonies. Apothecia rare; pycnidia common, dark red-orange, like tiny pimples. Lower surface white, sometimes with a few rhizines or holdfasts. Occasional, usually on bark, sometimes wood, rarely on rock, statewide, both coast and inland. Resembles *X. oregana,* which is lighter orange and has larger lobes and distinctive conidia. Photo from the Rocky Mountains, CO.

Xanthomendoza hasseana

Poplar sunburst lichen

Thallus bright orange, rather loosely attached, with divided, overlapping lobes. Apothecia numerous, with dark orange disks near the center of the thallus; apothecial margins often with tiny white rhizines underneath. Without soredia or isidia, but usually with numerous dark orange pycnidia. Lower surface white with long, white or yellowish rhizines. Spores ellipsoid, 15.5–18 × 7.5–9.5 µm with a broad septum. Fairly common on bark, occasional on wood or rock, statewide, Coast Range and inland. Resembles *X. montana*, which is a darker orange and more compact, with shorter, wider lobes and smaller, more cylindrical spores with a narrower septum less than one-third the length of the spore. Also resembles *Xanthoria polycarpa*, but that lacks true rhizines and the conidia differ. Photo from near Rifle, CO.

X. hasseana

Xanthomendoza montana

Mountain sunburst lichen

Thallus small, similar to *X. hasseana*, but often more compact (see discussion above). Spores 12–16 × 5–7.5 µm, septum 1.5–4 µm. Arid inland mountain areas in southern CA, less common than *X. hasseana*. Photo from the Mojave Desert.

Xanthomendoza oregana

Oregon sunburst lichen

Thallus small, foliose or subfruticose, with loosely adnate lobes that dissolve into blastidia at the margins, bearing granular soredia. Soredia can also grow on the lower surface, and the lobes sometimes roll inward, mimicking the hoods of *X. fallax*. Pycnidia with a mixture of ellipsoid and cylindrical conidia, a feature almost unique to this species. Lower surface white, smooth or a bit

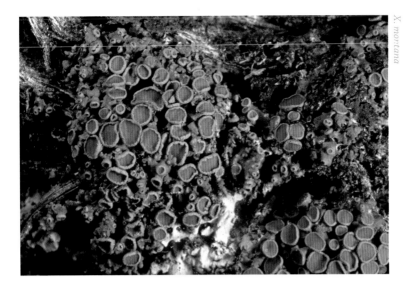

wrinkled, with white rhizines. Compare with *X. fulva;* also resembles *X. ulo-phyllodes,* but the lobes of *X. oregana* are narrower and more irregular, and the soralia are more frequent and never laminal. Statewide, mostly in the Coast Range and southern mountains. Photo from Mt. Palomar, San Diego Co.

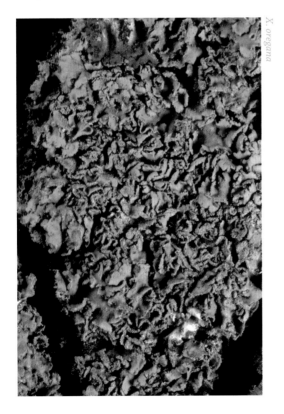

Xanthomendoza ulophyllodes
Powdery sunburst lichen

Thallus small, lobes radiating and branched, bearing marginal or laminal soralia with yellow-orange soredia; sometimes the soredia are spread over the surface. Apothecia uncommon; pycnidia frequent but not conspicuous. Lower surface white with rhizines. Can resemble *X. fallax,* but doesn't form hoods; *X. fulva* and *X. oregana* have soredia on the lower surface and more prominent pycnidia. Occasional on bark, probably statewide, more coastal than inland. Photo from coastal OR.

Also of Note
Another CA species, not described above, is *Xanthomendoza mendozae,* with a yellow to pale orange thallus forming centrally attached cushions, without apothecia, marginally sorediate and somewhat pruinose; on rock, but rare, in southern CA.

Xanthoparmelia
Rock-shield lichens

Small to large foliose lichens, yellowish green or somewhat bluish green; a few species, those formerly in the genus *Neofuscelia,* are dark olive to brown. More or less adnate, on siliceous rock or sometimes soil; very rarely on hard wood; some "vagrant" species are unattached, loose on soil. Many species have brown, lecanorine apothecia on the upper surface of the thallus; others have isidia, rarely soredia. Lower surface tan to black, usually with branched or forked rhizines. Spores ellipsoid, colorless, 8 per ascus. Spot tests: cortex usually K–, KC+ yellow; medulla various reactions. Common and widespread, often conspicuous, with many species found statewide, most often in drier habitats. Telling the species apart involves distinguishing the degree of attachment to the substrate, the presence and type of isidia, the color of the lower surface, and chemistry. A number of species not described below occur in CA, but they are relatively uncommon.

Xanthoparmelia coloradoënsis
Colorado rock-shield

Thallus adnate to fairly loose, 5–10 cm wide, fairly dark green, with irregular, somewhat overlapping lobes that often blacken at the margins, with loosely attached lobules on the surface. Without isidia, but usually with apothecia and black pycnidia. Lower surface tan with tan rhizines. Spot tests: medulla K+ yellow to red, KC–, C–, P+ orange. Fairly common on rock, statewide, especially in central to southern CA drier habitats and in the Sierra Nevada. Resembles *X. lineola* but more loosely attached, even becoming almost vagrant. Photo from interior BC.

X. coloradoënsis

Xanthoparmelia conspersa
Peppered rock-shield

Thallus pale yellow-green, forming rosettes 4–12 cm wide, appressed or loosely attached, with fairly narrow lobes 1–3 mm wide, often overlapping and brownish at the margins; with globular to branched cylindrical isidia and sometimes lobules in the central parts of the thallus. Apothecia uncommon, with isidiate margins. Lower surface black, except paler at the lobe tips, with black rhizines. Spot tests: medulla K+ deep yellow sometimes turning red, KC–, C–, P+ red-orange. Occasional, statewide. Resembles *X. plittii,* which has a tan to dark brown lower surface. Another similar but rather rare species from southern CA is *X. amableana* (not pictured), with an adnate thallus, lobes 0.8–2 mm wide, isidia initially spherical but becoming cylindrical or irregularly inflated, apothecia common, and a black lower surface; spot tests: upper cortex negative, medulla K+ yellow to orange, C–, KC–, P+ orange; the isidia in *X. amableana* are less consistently cylindrical than those of *X. conspersa.* Photo from the White Mountains, NH.

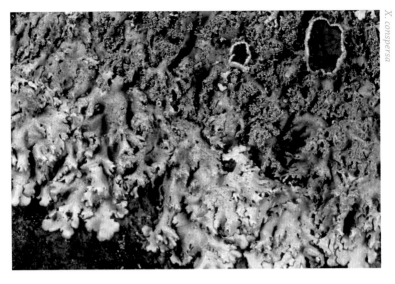

Xanthoparmelia cumberlandia

Cumberland rock-shield

One of the larger species, forming thalli up to 12 cm wide, pale green, closely or loosely appressed, with rounded or somewhat indented lobes, 1.5–4 mm wide, that are often black around the edges. Apothecia common, sometimes with in-rolled, toothed margins, 2–8 mm wide; pycnidia also frequent, forming patches of black dots on the surface; without isidia. Lower surface tan to brown with pale, simple rhizines. Spot tests: medulla K+ deep yellow slowly turning orange or red, KC–, C–, P+ orange, AI+ blue. Probably the most common species in CA, found on rock, occasionally on pebbles or soil, statewide, coast and inland. Photo from near Willits.

Xanthoparmelia lineola
Tight rock-shield

Thallus closely appressed, somewhat dull, yellow-green and often darker in the older parts, with rounded or irregular lobes 1–3 (occasionally up to 5) mm wide that sometimes overlap. Apothecia common, a bit raised; pycnidia abundant; without isidia. Lower surface tan, often wrinkled, with tan rhizines. Spot tests: medulla K+ blood red, KC–, C–, P+ deep yellow. Resembles *X. coloradoënsis* but more tightly attached. One of the more common species statewide, especially in central and southern drier habitats. Compare also with *X. wyomingica*. A rarer but very similar species from the southern interior is *X. subdecipiens* (not pictured); it is more loosely adnate and has negative spot tests in the medulla. Photo from Mt. Palomar, San Diego Co.

Xanthoparmelia loxodes
Blistered camouflage lichen

Thallus dark olive-brown or sometimes reddish brown with fairly long lobes; without true soredia, but with rounded pustules that look like globular isidia; these break down into schizidia, soredia-like fragments of the upper layers of the thallus. Apothecia uncommon. Lower surface dark brown to black, paler at the margins, with rhizines the same color as the lower surface. Spot tests: medulla K–, KC+ reddish violet to dingy orange, C– or yellowish, P–. This species, along with *X. subhosseana* and *X. verruculifera,* used to be in the genus *Neofuscelia;* they resemble *Melanelia* species in form and color but lack pseudocyphellae and differ in chemistry. *X. loxodes* resembles *X. verruculifera,* which is smaller and darker, with narrower, thinner lobes that usually react KC– or faintly reddish in the medulla. Uncommon on rock in dry habitats statewide. Photo from southern interior BC.

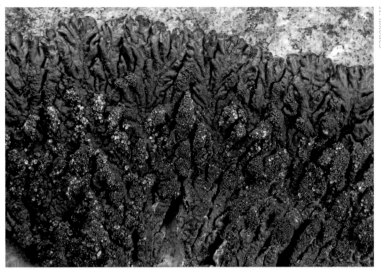

Xanthoparmelia maricopensis
Maricopa rock-shield

Thallus yellow-green, adnate, with irregular, somewhat shiny lobes, 1–2 mm wide, that are moderately isidiate; isidia are spherical, becoming cylindrical to irregularly inflated and branched. Lower surface tan with tan, unbranched rhizines. Spot tests: medulla K+ yellow to orange, KC–, C–, P+ orange. On rock in southern deserts. Similar to the more common *X. mexicana* but differs in chemistry. A rarer species from southern deserts, *X. dierythra* (not pictured), has larger, flatter lobes and a K+ red medulla. Photo from Tonto National Forest, AZ.

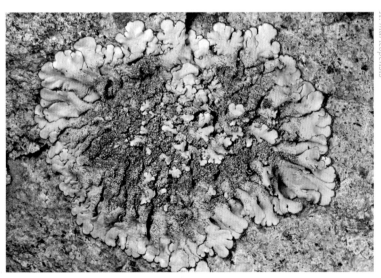

Xanthoparmelia mexicana
Salted rock-shield

Thallus pale yellow-green, loosely or tightly appressed, with rounded lobes, 1.5–5 mm wide, that are sometimes shiny at the tips; usually with abundant isidia that are at first spherical then cylindrical and branched. Apothecia and pycnidia rare. Lower surface pale brown, sometimes grayish at the margins, with pale brown rhizines. Spot tests: medulla K+ yellow turning dark red, KC–, C–, P+ orange. Common statewide, most often in central to southern arid habitats but can be coastal as well. Compare with *X. maricopensis.* Photo from the Channel Is.

X. mexicana

Xanthoparmelia mougeotii
Mougeot's rock-shield

Thallus very small, sometimes appearing areolate in the center, tightly appressed and very irregular in shape, with lobes only 0.2–0.5 mm wide, pale to dark yellow-green and partially brown, with farinose soredia in capitate soralia. Apothecia very rare. Lower surface black, shiny, moderately rhizinate. Spot tests: medulla K+ yellow turning dark red, KC–, C–, P+ orange. Unusual in its tiny size and presence of soredia. Uncommon, usually on rock, rarely on tree bases, in parts of the Coast Range with some maritime influence, very occasionally on the western slope of the Sierra Nevada. Photo from Mt. Tamalpais, Marin Co.

Xanthoparmelia novomexicana

New Mexican rock-shield

Thallus yellow-green, closely attached, forming rosettes 4–20 cm wide, with irregular lobes that are smooth but become rugose as they age, 1–3 mm wide; without isidia. Apothecia common, round, almost black, 1–4 mm wide. Lower surface and rhizines pale brown. Spot tests: medulla K– or K± brownish, KC–, C–, P+ red. Fairly common on rock in central and southern CA mountain regions and the Sierra Nevada. Photo from Big Bend National Park, TX.

Xanthoparmelia plittii

Plitt's rock-shield

Thallus yellowish green with abundant isidia in the central areas that are initially spherical but become cylindrical and branched. Almost identical to *X. conspersa*, except the lower surface is pale to dark brown, never black, and sometimes mottled. Spot tests: medulla K+ deep yellow turning orange, KC–, C–, P+ red-orange. Another fairly common species in both coast and interior is *X. subplittii* (not pictured), which differs in having narrower lobes and spherical isidia. Statewide, in drier habitats. Photo from southern interior BC.

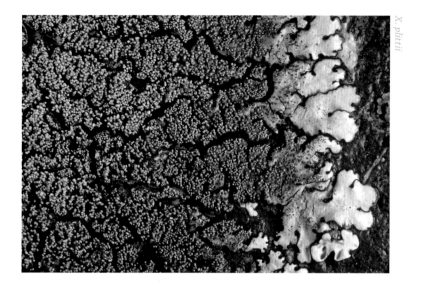

Xanthoparmelia subhosseana
Erupted camouflage lichen

Thallus grayish or slightly reddish olive-brown, resembling *X. loxodes* but closer to *X. verruculifera,* with disintegrating pustules of isidiate fragments that make it look sorediate. Lower surface dark brown to black, with rhizines the same color as the lower surface. Spot tests: medulla K+ deep yellow becoming red, KC–, C–, P+ orange but slowly. Uncommon on rock in drier sites statewide. Photo from Boise, ID.

Xanthoparmelia subramigera

Branching rock-shield

Thallus yellow-green, with narrow, divided lobes, 1–4 mm wide, moderately to densely isidiate, quite similar to *X. mexicana* except for medullary chemistry. Spot tests: medulla K–, KC–, C–, P+ orange or red. Moderately common in southern CA near the coast and in the Sierra Nevada. Photo from Pedernales Falls State Park, TX.

X. subramigera

Xanthoparmelia verruculifera

Warty camouflage lichen

Thallus grayish olive-brown, similar to *X. loxodes,* with abundant granular isidia that form patches on the surface, soon breaking down into schizidia. Apothecia rare. Spot tests: cortex negative except HNO_3+ dark blue-green; medulla K–, KC–, or KC+ reddish, C– or C+ rose, P–. Fairly common on rock,

X. verruculifera

statewide, mostly in drier inland sites but can also be coastal. Photo from Boise, ID.

Xanthoparmelia wyomingica
Shingled rock-shield, Variable rockfrog
Thallus pale green, loosely attached or almost vagrant, forming little piles of overlapping, narrow, irregular lobes that often have black edges. Lower surface black near the lobe tips but pale in the center, with sparse rhizines. Spot tests: medulla K+ yellow turning red, KC–, C–, P+ yellow. Rather similar to *X. lineola* but much more loosely attached; *X. coloradoënsis* is intermediate between the two. Uncommon on pebbles, soil, or rock, sometimes at higher elevations but coastal in southern CA. Photo from southern interior BC.

Xanthoria
Sunburst lichens, Orange lichens
Small, bright orange, foliose lichens, often sorediate, with lecanorine apothecia. Spores polarilocular, 8 per ascus. Spot tests: upper cortex K+ purple, KC–, C–, P–. On bark, wood, and rock, especially when calcareous. Until recently, the genus included species now within *Xanthomendoza,* but *Xanthoria* species lack true rhizines and have different conidia.

Xanthoria candelaria
Shrubby sunburst lichen
Thallus yellow-orange with very small, divided, ascending lobes that give it a fruticose appearance, forming cushions up to about 3 cm wide or spreading into larger colonies. Usually with granular blastidia or soredia on or close to the margins. Apothecia occasional, darker orange. Lower surface white or yellowish. Fairly common on rock or bark statewide, usually in coastal areas, but occasionally inland as well, such as in the Sierra Nevada. Often grows on

rocks where birds leave nitrogen-rich deposits. A similar species is *X. ascendens* (not pictured), which has hooded lobe ends with powdery soredia; uncommon on the Channel Is. When fertile, *X. candelaria* might be mistaken for *Caloplaca coralloides,* which is not sorediate, is more truly fruticose, and is almost always found near the ocean. Photo from the Kenai Peninsula, AK.

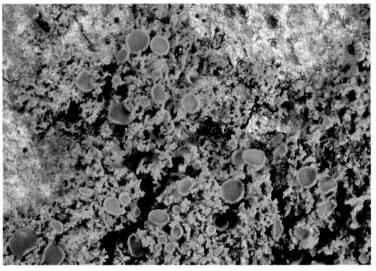

X. candelaria

Xanthoria elegans

Elegant sunburst lichen

Thallus bright yellow-orange to red-orange, up to 6 cm wide, forming rosettes with radiating lobes, closely appressed to the substrate and almost crustose. Apothecia usually abundant. Without soredia or isidia but sometimes with papillate growths on the surface and slightly pruinose. Lower surface white

X. elegans

and wrinkled, without rhizines. Can resemble the brighter red-orange species of *Caloplaca,* such as *C. ignea* or *C. saxicola,* but they are true crustose lichens that lack a lower cortex and cannot be peeled from the substrate. Quite common and conspicuous on rock in drier inland habitats statewide, sometimes coastal as well. Photo from Inyo Co., alpine Sierra Nevada.

Xanthoria parietina
Maritime sunburst lichen

Thallus small to medium-sized, yellowish orange or orange, sometimes greenish when in the shade, forming rosettes up to 10 cm wide, but usually smaller, with broad, radiating, wrinkled lobes 0.7–3.5 mm wide, fairly tightly appressed. Apothecia common, darker orange, especially near the center. Lower surface white, attached to the substrate with holdfasts or with a few short rhizines. On bark, especially on California buckeye (*Aesculus californica*), or on wood or rock, statewide, mostly in coastal habitats. Photo from near Santa Cruz.

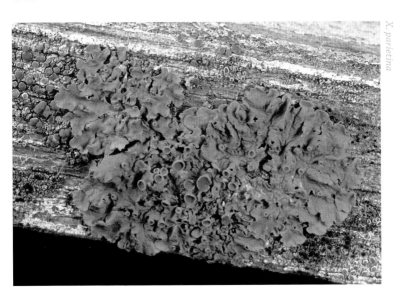

Xanthoria pollinarioides
Pollen sunburst lichen

Thallus small, compact, almost fruticose, orange, up to 1 cm long, with powdery soredia that emerge in rounded ruptures or on the lobe tips. Apothecia rare. Lower surface paler, with some soredia. Apothecia rare. Somewhat resembles a tiny version of *Teloschistes*. Uncommon on coastal shrubs and trees such as *Baccharis* and *Salix,* Marin Co. to Mexico; easily overlooked because of its small size. Photo from Bodega Bay, Sonoma Co.

Xanthoria polycarpa
Pin-cushion sunburst lichen

Thallus forming small, raised, orange cushions up to 2.5 cm wide, with very narrow, divided and irregular lobes. Apothecia abundant, covering the central portion. Without soredia or isidia. Lower surface white, sometimes with broad

holdfasts but without true rhizines. Pycnidia usually present, immersed, containing ellipsoid conidia. Common on bark, usually twigs, in fairly humid sites, rarely on rock, statewide, most often in the Coast Range but inland as well. Resembles *Xanthomendoza hasseana* and *X. montana,* but those have rhizines and rod-shaped conidia. Photo from a gravestone, coastal OR.

Xanthoria sorediata
Sugared sunburst lichen

Thallus with radiating orange lobes, rather like *X. elegans,* but the older parts of the lobes have spherical pustules that break down into coarse, granular soredia. Apothecia rare. Occasional on rock, especially calcareous sandstone or nutrient-rich surfaces, in dry, inland locations, central and southern CA and in the Sierra Nevada. Photo from southern ON.

Also of Note
An additional CA species is *Xanthoria tenax* (not pictured), which has a tightly adnate, often lightly pruinose thallus forming rosettes up to 2.5 cm wide; only the lobe tips have a lower cortex; apothecia common, laminal; rather common on bark, often on twigs, most often in habitats with some coastal fog, central and southern Coast Range.

Fruticose Lichens

--- --- --- --- --- --- --- --- ---

Alectoria

Witch's hair

Pendant or shrubby fruticose lichens, pale yellowish green, with slender branches that have raised white pseudocyphellae; branches usually rounded but can be somewhat flattened or pitted. Spores large, brownish, 1-celled, 2–4 per ascus. *Alectoria* species are common and widespread from OR to AK, occurring only in northern CA. Can be mistaken for *Usnea;* if you stretch a branch and it breaks cleanly, it is *Alectoria;* if it reveals a somewhat elastic central cord, it's a species of *Usnea*.

Alectoria imshaugii

Spiny witch's hair

Thallus shrubby or almost pendant, 5–8 cm long, with isidia along the branches growing from fissures in the cortex. Spot tests: cortex and medulla K+ yellow, KC–, C–, P+ orange; or all tests negative. Looks much like an *Usnea* species but lacks a central cord. Uncommon on conifers in northwestern CA. Photo from the Oregon Dunes.

Alectoria lata

Flowering witch's hair

Thallus shrubby or somewhat pendant, typically 5–8 cm long, occasionally longer. Apothecia usually abundant, dark brown, lecanorine; without soredia or isidia. Spot tests: cortex K–, KC+ gold, C–, P–; medulla usually KC+ red. Uncommon on conifers, especially pine, in northwestern CA. Photo from Horse Mountain, east of Eureka.

Alectoria sarmentosa

Witch's hair

Thallus pendant, pale yellow-green, easily mistaken for some species of *Usnea* but without a central cord; the hairlike strands can reach 40 cm or more in length. Apothecia sometimes present, light brown, lecanorine. Spot tests: cortex K–, KC+ gold, C–, P–; medulla C–, KC+ red. Common on conifers in

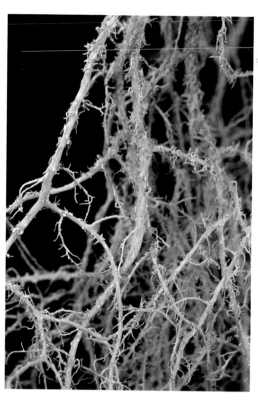

A. imshaugii

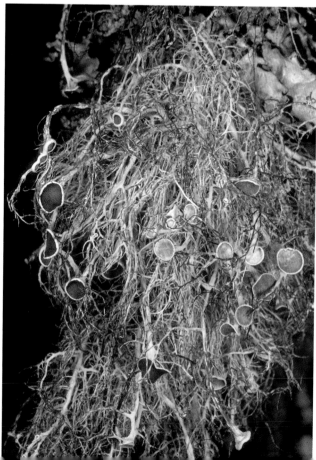

A. lata

mountains from OR to coastal AK, but occasional in northern CA. A rarer species in coastal northern CA, *A. vancouverensis* (not pictured), is coarser and slightly grayer; spot tests: medulla C+ and KC+ deep red. Photo from Wenatchee National Forest, eastern WA.

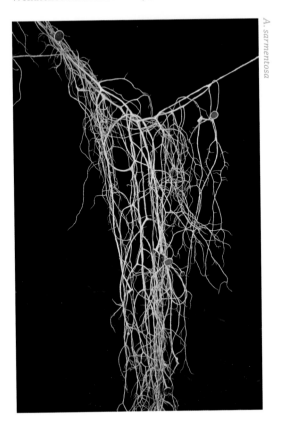

A. sarmentosa

Bryoria
Horsehair lichens, Tree-hair lichens, Bear hair
Fruticose, hairlike lichens, mostly pendant but some species more shrubby-erect; usually brown but a few species pale and some forms are deep yellow. Apothecia uncommon in most species, but soredia, isidia, or pseudocyphellae occur in many. Spores are ellipsoid, colorless, 1-celled, 8 per ascus. Chemical tests require the filter paper method. Most commonly seen on trees, especially conifers, rarely on rock. *Bryoria* species look like a brown or dying *Usnea* but lack *Usnea*'s central cord. Compare to *Nodobryoria, Pseudephebe, Sulcaria badia*, possibly *Kaernefeltia californica*.

Bryoria capillaris
Gray horsehair lichen
Thallus usually pale gray but can be smoky brown, with slender strands, 0.1–0.3 mm thick, up to 30 cm long, fairly brittle, with short or long, narrow, pale pseudocyphellae appearing like a pale line running down the branches,

sometimes in a spiral; without soredia. Spot tests: cortex and medulla K+ yellow, KC+ red, C+ pink, P+ deep yellow. On conifers in northwestern CA. Photo from Mt. Hood National Forest, OR.

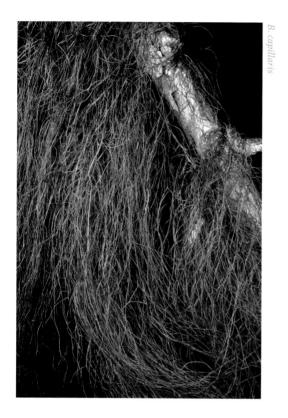

B. capillaris

Bryoria fremontii

Tree-hair lichen, Black tree-lichen, Edible horsehair

Thallus pendant, shiny, red-brown to yellow-brown, with thick and twisted branches 10–30 cm long, uneven with dents and ridges and short, perpendicular side branches; sometimes with bright yellow soralia; not brittle. Apothecia occasional, brown, but with bright yellow pruina. Spot tests negative. Fairly common on conifers in drier habitats in northeastern mountain areas, but occurs throughout most of CA. The lichen formerly called *B. tortuosa,* with a yellowish brown to dark yellow thallus, twisted and pitted main branches with slender, perpendicular side branches, and often with long, yellow, spiraling pseudocyphellae, is now included in *B. fremontii*. Indigenous people in BC cooked *B. fremontii* in pits, mixed it with honey, and ate it, one of the few known traditional uses of lichens as food; the yellower form, however, can have high concentrations of vulpinic acid, which is poisonous. Compare with *Nodobryoria oregana*. Photo from near Canby, Modoc Co.

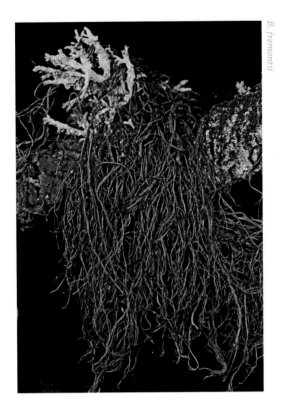

Bryoria furcellata

Burred horsehair lichen

Thallus bushy, shiny, dark brown, in clumps up to 5 cm across with pointed side branches that make it look thorny. Numerous soralia in cracks contain white farinose soredia, together with short, spiny isidia; without pseudocyphellae, but young, undeveloped soralia can look like them. Spot tests: cortex and medulla K–, KC–, C–, P+ red. The isidia growing out of fissural soralia

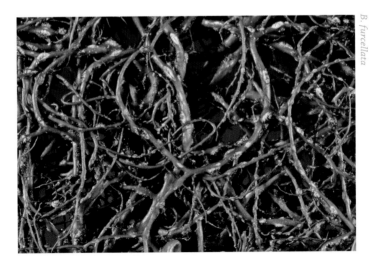

make the species distinctive. Uncommon on conifers in open forests, north-western CA. Photo from coastal OR.

Bryoria fuscescens
Pale-footed horsehair lichen

Thallus pale to very dark brown or olive-brown, strands 5–15 cm long, with the base often paler than the rest, commonly with white, fissural or tuberculate soralia and occasional spiny side branches. Spot tests: cortex and medulla usually negative but sometimes P+ red; soralia P+ red. Occasional on conifers, northwestern CA. Compare with *B. glabra*. A similar but rare species, *B. subcana* (not pictured), is pale brown to cream; spot tests: thallus P+ red. Photo from interior AK.

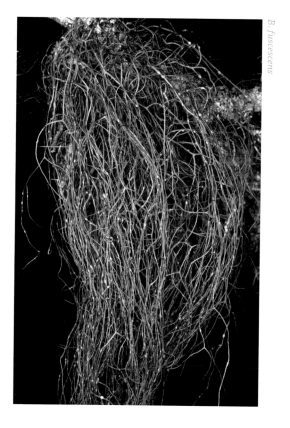

B. fuscescens

Bryoria glabra
Shiny horsehair lichen

Thallus pendant, greenish brown or olive-brown and shiny, in clumps 10–15 cm long, with uniform branches 0.2–0.4 mm thick; branching in wide, equal dichotomies, not twisted. Abundant or sparse, pure white, elongated (fissural) soralia; without spiny side branches or pseudocyphellae. Rather similar to *B. fuscescens*, but that has branches that fork less evenly and at a more acute angle, and they are sometimes twisted and flattened at the axils. In addition, it often has tuberculate rather than elongate soralia. Spot tests: cortex and medulla negative, but soralia P+ red. Uncommon on conifers at all elevations, northwestern CA. Photo from Olympic National Park, WA.

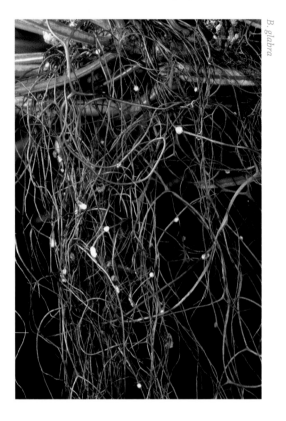

B. glabra

Bryoria pseudofuscescens
Mountain horsehair lichen

Thallus pendant, pale gray-brown to almost black, strands 5–10 cm long and 0.15–0.8 mm thick; branches slender, uneven, and twisted. Without soralia but usually with abundant, white, fusiform to elongated pseudocyphellae that spiral around the branches. Apothecia rare. Spot tests: cortex and medulla K+ yellow, changing to dull red with the filter paper method, KC–, C, usually P+ yellow. On conifers in montane or subalpine forests, sometimes almost covering whole trees. In many areas a significant food source for elk and deer, and in some habitats provides food and nesting material for northern flying squirrels. Photo from eastern OR.

Bryoria spiralifera
Spiral horsehair lichen

Branches variable in color but mostly pale red-brown with numerous spiny side branches and often with long pseudocyphellae that typically spiral. Spot tests: K+ red, KC–, P+ yellow. Rare, concentrated on conifers in coastal dunes,

central to northern CA. A similar, probably even rarer species that shares the same sort of marine habitat is *B. pseudocapillaris* (not pictured); it is more uniformly pale with linear pseudocyphellae; spot tests: K+ yellow, KC+ red, P+ yellow. Photo from the coast near Eureka.

Bryoria trichodes subsp. *americana*

American horsehair lichen

Thallus pendant, pale to dark brown, with even, smooth branches 7–15 cm long, with short, slender, slightly depressed pseudocyphellae. Apothecia common, reddish brown. Spot tests: outer cortex negative; medulla and inner cortex K–, KC–, C–, P+ red, seen with the filter paper method. Recent work indicates that the taxon should not be considered a subspecies of *B. trichodes* but should be called *Bryoria americana* instead. Uncommon in coastal forests, central and northern CA. Photo from Southeast AK.

B. trichodes

Cladonia

Cladonia

Most species have a squamulose primary thallus, and most develop hollow erect stalks called podetia, which may be simple or branched. Because most species of *Cladonia* are dominated by their podetia, the genus is grouped here with the fruticose lichens. In many species the tips of the podetia have biatorine apothecia of various shapes, usually red or brown. Podetia have a cortex, with a layer of algae beneath, and a blackened, interior supporting layer called a stereome, partly visible in some species. Species with multiple branches and lacking a cortex used to be classified in the genus *Cladina*, but they are now included within *Cladonia*; with these, the branches arise from an inconspicuous granular crustose thallus, rather than from basal squamules. Spores are

C. asahinae

colorless, 1-celled, 8 per ascus. Widespread, usually on soil or decaying wood, but also on bark or rock, most often in shady habitats, but occasionally in open areas; they often colonize disturbed habitats. *Cladonia* species are generally not as common in CA as they are in the more northern and eastern parts of North America.

Cladonia asahinae
Western pixie-cup

Primary thallus squamulose, sometimes sparse, gray-green to slightly yellowish or whitish. Podetia cupped, toothed, 10–25 mm tall, sometimes proliferating at the margins. Upper part of podetia with corticate granules or granular to powdery soredia. Apothecia uncommon, brown. Part of the "*chlorophaea* complex." Spot tests: podetia P+ red, K–, KC–, C–, UV–. Resembles *C. fimbriata*, which is less gray and has narrower, more uniform cups covered in fine powdery soredia. The two species also differ chemically (*C. asahinae* contains fatty acids), though not in spot tests. Occasional on soil and mossy rock statewide. Compare with *C. chlorophaea* and other cupped species. Photo from Monterey Co.

Cladonia bellidiflora
Toy soldiers

Thallus pale yellow-green with deeply lobed, large squamules, without soredia; podetia covered with squamules. Apothecia conspicuous, bright red. The combination of large, squamulose podetia without soredia and showy red apothecia on top is distinctive. Spot tests: cortex KC+ gold, most specimens K–, P–, UV+ blue-white, occasionally K+ deep yellow, P+ orange. A Pacific Northwest species. On moss and rotting wood, northwestern CA. Photo from Mt. Hood National Forest, OR.

C. bellidiflora

Cladonia cariosa
Split-peg lichen, Split-peg soldiers
Primary thallus of thick, gray-green, persistent squamules that can be tongue-shaped, toothed, or lobed. Podetia pale gray to greenish, 5–35 mm tall, often branched once or twice, fissured, rough-looking, without cups or soredia. Apothecia large, brown, on the tips of the podetia. Spot tests: K+ yellow, sometimes becoming brownish, C–, KC–, P+ yellow, or sometimes becoming red, or P–, UV–. Occasional on soil in damp habitats statewide. Photo from Olympic National Forest, WA.

Cladonia carneola
Crowned pixie-cup
Primary thallus of yellow-green, deeply divided squamules, without soredia or sometimes with some on lower surface near the margins. Podetia forming cups with jagged margins, covered with powdery soredia both inside and (usually) outside the cups. Apothecia frequent, small, pale brown; pycnidia pale pink or brown. Spot tests: surface K–, KC± gold, usually C–, P–, UV–.

Occasional on rotting wood, soil, and bark, mostly in northeastern CA. The yellowish tint and powdery soredia, together with the toothed cups, make the species distinctive. Compare with *C. chlorophaea* and other cupped species. Photo from Olympic National Forest, WA.

Cladonia chlorophaea
Mealy pixie-cup
Primary thallus of deeply lobed squamules, without soredia. Podetia pale gray-green, sometimes slightly brownish, forming broad cups, 2–6 mm wide, with granular soredia both inside and outside the cups. Squamules sometimes form on the outside of the podetia, but sparsely, and they are small, almost granular. Podetia can proliferate and may have large brown apothecia on the margins of the cups, but more often without apothecia. Spot tests: K–, KC–, C–, P+ red. A common, widely distributed species, and variable; can be confused with *C. asahinea, C. carneola* (usually more yellowish), *C. fimbriata,* or *C. pyxidata.* Also compare with *C. humilis* and *C. nashii,* listed in "Also of Note," below. Photo from eastern ID.

C. chlorophaea

Cladonia concinna
Slender ladder lichen, Ballerina lichen
Basal squamules sparse and inconspicuous. Podetia pale gray-green to slightly brownish, without soredia, smooth or verruculose, with multiple, slender, stalked cups, each one growing from the center of the cup below, often forming tiers several cups high. Spot tests: thallus K–, KC–, C–, P+ red. Unmistakable but uncommon on thin, often acidic, soil in open areas or bogs in coastal locations, central and northern CA. A similar species, *C. verticillata* (not pictured), has shorter and broader cups and is more northern, probably not found in CA. Photo from western OR.

Cladonia coniocraea
Common powderhorn
Primary thallus with large, green to olive squamules up to 6 mm wide, rounded or divided. Podetia pale green, slender, tall (10–25 mm), often tapering and usually unbranched, growing mostly from the centers of the squamules, covered with powdery soredia, rarely with small brown apothecia.

C. concinna

C. coniocraea

Infrequently, tiny cups form on the tips of podetia. Spot tests: thallus K– or brownish, KC–, C–, P+ red. Occasional on soil, usually in the shade, on tree bases, sometimes on wood; statewide, but more coastal and northern. Compare with *C. ochrochlora,* which has squamules that are nearly entire, grayer podetia that are more commonly cupped, and patchier, more granular soredia. Photo from Monterey Co.

Cladonia fimbriata

Trumpet lichen

Primary thallus of persistent, greenish to olive-gray, lobed squamules, usually without soredia but sometimes coarsely sorediate under the margins. Podetia greenish to pale gray, narrow, 10–20 mm tall, covered with fine, farinose, occasionally granulose, soredia, with symmetrical cups that have even margins. Apothecia uncommon, brown. Spot tests: thallus K– or brownish, KC–, C–, P+ red. Podetia more slender than *C. chlorophaea;* greener and more powdery than *C. asahinea.* On soil and rotting wood, more often in mountains than along the coast, central to northern CA. Photo from southern BC.

Cladonia firma
Alternation Cladonia

Primary thallus small, squamulose, with deeply lobed squamules up to 10 mm wide and 10 mm long, white to purplish gray below, often with white tips, without soredia. Occasionally with small, cupped podetia that have corticate granules and squamules on the surface and cup margins. The lobes of the thallus curl up when the lichen is dry, showing the lower surface; when wet, it uncurls and appears much greener, a characteristic similar to *C. prostrata* (not pictured), which grows on sand in Florida. It often lacks podetia, making it inconspicuous. Spot tests: K+ yellow, C–, KC–, P+ red, UV–. More common in Europe, it is very rare in CA, known from only a few sandy localities in coastal San Luis Obispo Co. and is state-listed as an endangered species. Compare with *C. macrophyllodes*.

Cladonia firma in the dry state.

Cladonia firma when damp. Photos from San Luis Obispo Co.

Cladonia furcata
Many-forked Cladonia

Primary thallus with greenish gray to brown squamules that disappear as the lichen matures. Podetia greenish or bluish gray, often brown in sunny locations, slender, forking into multiple branches, forming clumps 2–12 cm tall. Branches sometimes with podetial squamules. Axils open, but without cups. Apothecia often on tips, brown, up to 1.5 mm wide; frequently also with pycnidia. Spot tests: K–, or K+ dingy yellow to brownish, C–, KC–, P+ red, UV–. Occasional on soil and mossy rocks in shady coastal forests, entire length of coast. Intergrades with *C. scabriuscula,* which typically has fewer branches with granular soredia on the tips. Photo from western WA.

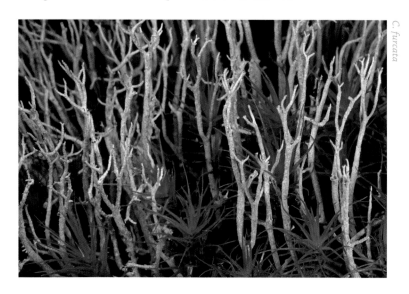

C. furcata

Cladonia macilenta

Lipstick powderhorn, pin lichen

Primary squamules small, pale gray to light green, 1–6 mm long, sometimes with granular soredia. Podetia slightly lighter in color, slender, unbranched, 10–30 mm tall, blunt at the tips but without cups. Apothecia frequent on top,

C. macilenta

small, bright red. There are two chemotypes: one is K–, KC+ yellow to orange, C–, P–, recognized as variety *bacillaris;* the other is K+ yellow, KC–, C–, P+ orange, which is variety *macilenta.* On wood, bark, especially around tree bases, soil, or occasionally on mossy rock, in coastal habitats statewide. One of the more common CA species of *Cladonia,* noticeable because of the red-tipped podetia, even though small in size. Photo from western OR.

Cladonia macrophyllodes
Large-leaved Cladonia
Primary thallus persistent with large lobed squamules, up to 8 mm wide and 15 mm long, pale gray-green or brownish, often with waxy-looking small bumps. Podetia short, cupped, sometimes pruinose, proliferating from the centers of basal lobes, often with small squamules along the margins of the cups. Apothecia occasional, small, brown. Lower surface white, without a cortex. Spot tests: P+ red, K+ yellow, KC–, C–, UV–. A lichen of mountain habitats, uncommon in northern CA, absent farther south. *C. firma* also has very large squamules, but they lack whitish bumps, and the lower surface usually shows some purplish gray pigmentation. Photo from western CO.

C. macrophyllodes

Cladonia ochrochlora
Smooth-footed powderhorn
Primary thallus with large, green, lobed squamules. Podetia unbranched, tall (1–3.5 cm), often tapering, frequently with narrow cups, usually with a cortex on the lower half and mealy soredia on the upper half. Apothecia often on tips, pale red-brown. Spot tests: K–or K+ dingy yellow to brownish, KC–, C–, P+ red, UV–. On old wood, particularly of conifers, and frequent on fences, in coastal habitats statewide. Compare with *C. coniocraea.* Photo from Olympic National Forest, WA.

C. ochrochlora

Cladonia portentosa subsp. *pacifica*

Maritime reindeer lichen

Primary thallus crustose and almost always absent; shrubby growth form with erect, thin, many-branched podetia, branching mostly in threes or sometimes twos, very pale greenish to white, often browning at the tips. Branches are without an outer cortex. Spot tests: K–, KC+ yellow, or KC–, C–, P–, UV+ blue-white. Rare on mossy or sandy soil in oceanic habitats, Mendocino Co. north. Photo from the Pygmy Forest near Fort Bragg, Mendocino Co.

Cladonia pyxidata

Pebbled pixie-cup

Primary squamules thick, persistent, tongue-shaped, with pale gray-green to olive-brown, mostly ascending lobes. Podetia short, cupped, with rounded, corticate areoles or flattened squamules on the inside of the cups, sometimes also on the outside of the podetia. Surface of the podetia can be rough, appearing verruculose or granular. Apothecia frequent on the cup margins, small, brown, creating the appearance of a thin brown ring. Spot tests: thallus K– or brownish, KC–, C–, P+ red. On soil, especially if acidic, coast and mountains statewide. One of the more common species of *Cladonia* in CA. Closely resembles *C. chlorophaea;* the two species can be hard to distinguish. Photo from Humboldt Co.

Cladonia scabriuscula
Mealy forked Cladonia

Primary squamules rare. Podetia pale grayish-green to brown, slender, as tall as 10 cm, branching near the top, usually with open axils and with granular soredia at the tips; tips sometimes also with scurfy bits of nonsquamulose cortex. Apothecia uncommon, brown. Spot tests: K–, or K+ dingy yellow to

brownish, KC–, C–, P+ red, UV–. Occasional in western CA. Compare with *C. furcata*. Photo from southern ON.

Cladonia squamosa
Dragon Cladonia, Dragon funnel

Thallus with abundant and persistent primary squamules, finely divided, without soredia. Podetia greenish to very pale gray, or brownish, covered

with squamules but with a disintegrating cortex, often branched at the top, sometimes with small cups. Two chemical races: podetia K–, KC–, C–, P–, UV+ blue-white; or K+ deep yellow, KC–, C–, P+ deep yellow-orange. Occasional on soil, decaying wood, or mossy rock in moist habitats, mostly in northern CA. Photo from Ozark National Forest, AR.

Cladonia subulata
Antlered powderhorn
Primary squamules sparse to disappearing. Podetia pale greenish gray, slender, tall (3–10 cm), covered with powdery soredia their entire length, sometimes spotted with brown. Often with narrow cups at the tips or one or two stubby short branches that branch at a wide angle. Spot tests: K– or brownish, KC–, C–, P+ red, UV–. Uncommon on soil or rotting wood statewide. Photo from interior AK.

Cladonia verruculosa
Western wand lichen, Pebblehorn lichen
Primary squamules small to absent. Podetia pale brown, cream-colored or olive, slender and tall, to 7 cm; tips may be pointed or have small cups; granular soredia cover the upper half of podetia or the entire length. Podetia surface without a cortex, revealing the brown to black horny stereome near the base. Apothecia frequent, brown, up to 3 mm wide. Spot tests: K–, or K+ dingy yellow to brownish, KC–, C–, P+ red. Occasional, mostly in sunny habitats, sometimes on sand dunes, statewide. Photo from Olympic National Forest, WA.

C. verruculosa

Also of Note

Some additional species, not pictured here, occur in CA. The more common ones include:

Cladonia hammeri, with a squamulose primary thallus and cupped, granulose podetia 4–10 mm tall; apothecia rare, brown; spot tests: K–, C–, KC–, P+ red, UV–; on soil along the central and southern coast and in coastal mountains.

Cladonia humilis, which resembles *C. chlorophaea,* with stout podetia typically 5–12 mm tall and wide cups; cortex continuous on the lower part of each podetium; upper part with farinose soredia; apothecia very rare; spot tests: K+ yellow becoming dingy yellow, C–, KC–, P+ brick red, UV–; central and southern coast and in coastal mountains.

Cladonia nashii, which is very similar to *C. humilis,* above, but without the smooth cortex, with coarser soredia, and differing chemically; apothecia absent; spot tests: K+ yellow, C–, KC–, P+ red, UV–; fairly common in the Coast Range, Monterey Co. to Mexico.

Cladonia pocillum, with thick, brownish squamules in a flat mat, looking almost like a foliose lichen; podetia typically short, brownish, without soredia, with round areoles inside and outside the cups like *P. pyxidata;* apothecia brown, on cup margins; spot tests: thallus K– or brownish, KC–, P+ red; occasional in montane areas, central and southern CA.

Cladonia subfimbriata, with persistent primary squamules and cupped, whitish gray podetia 4–15 mm tall with a rough, sorediate surface; apothecia rare; spot tests: thallus K+ reddish brown, C–, KC–, P+ red, UV–; somewhat intermediate between *C. fimbriata* and *C. subulata;* in southern coastal ranges.

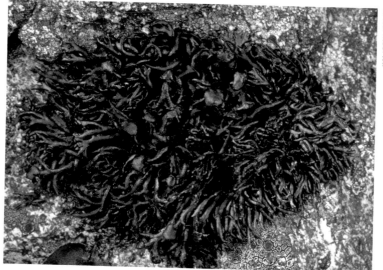

Cornicularia normoerica
Bootstrap lichen

A small, black to dark red-brown, fruticose lichen with narrow flattened lobes up to 0.5 mm wide that branch dichotomously, often with lecanorine apothecia near the branch tips. The cortex is thick and tough, the medulla dense. Spores ellipsoid, colorless, 1-celled, 8 per ascus. Spot tests negative. Occasional on exposed rock in northern mountains. Species of *Pseudephebe* grow in similar habitats and look somewhat similar but have finer, more complicated branches that are more woolly than straplike. *Lichinella* species have a cyanobacterial photobiont and look more gelatinous when wet. Photo from Siskiyou National Forest.

Dendrographa
False orchil

The most commonly encountered species have shrubby fruticose thalli, usually whitish or pale gray but occasionally darker, attached to the substrate with a single holdfast and with flattened or rounded (terete) branches, 1–9 cm long. The surface often looks bumpy due to short, white tubercules that push through the cortex. Branches sometimes have heavily pruinose fruiting bodies (ascomata) that resemble lecanorine apothecia but actually have a different type of development and structure. *D. conformis* is crustose and *D. franciscana* is lumpy. Spores are fusiform, colorless, 4-celled, 8 per ascus. Spot tests: cortex K– or yellowish brown, C–, KC–, P+ orange to red; medulla K– or pale yellow, C–, KC+ pink, P+ orange to red. Fruticose species of *Dendrographa* resemble the much rarer, true orchil lichen, genus *Roccella*, which reacts C+ red, P–. *Dendrographa* grows on rocks, wood, and bark in the fog zone near the ocean, central to southern CA.

Dendrographa alectoroides

Slender false orchil

Thallus fruticose, with pale gray terete branches, usually less than 3 cm long. Can have small ascomata and black pycnidia; typically coated with a thin white pruina. The photo shows form *parva*, which is found from southern CA to Marin Co. The somewhat less common form *alectoroides*, with flattened main branches, has a more limited range, found mainly from Marin to Monterey Co.; it can be confused with *D. leucophaea*, but the small, lateral branches are round in cross section, not flattened. Photo from Pt. Lobos, Monterey Co.

Dendrographa conformis

Peg-legged lichen

Thallus crustose, grayish tan to slightly greenish, rather rough, sometimes with a darker prothallus. Apothecia lecanorine, black but with heavy white pruina; without soredia. Spores banana-shaped (somewhat curved), 26–31 × 4–5 μm. Spot tests: K– or yellow turning red, KC–, C–, P+ orange. Occasional on bark of trees and shrubs, sometimes on wood or rock, in very coastal sites, San Luis Obispo Co. to southern CA. Resembles *Dirina paradoxa* subspecies *aproximata* (not pictured), which reacts C+ red and occurs in about the same localities. Photo from the Channel Is.

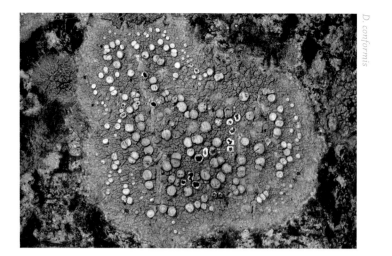

Dendrographa franciscana
Chalky peg-leg lichen

Thallus bullate, often forming cushions, whitish. Apothecia somewhat immersed, irregularly shaped, black. Spores banana-shaped (somewhat curved), 26–31 × 4–5 µm. Spot tests negative. Variable; some forms can resemble the rarer *Schizopelte crustosa* (not pictured). On bark and wood, occasionally on rock, in coastal habitats, San Francisco to Mexico. Photo from Pt. Lobos State Park, Monterey Co.

Dendrographa leucophaea

Robust false orchil

Shrubby, with mostly flattened branches, especially near the axils, pale gray to dark yellowish gray. Can have small ascomata, also black pycnidia, and usually has a coating of thin white pruina. *D. leucophaea* (without any designated form) and form *minor,* which has shorter internodes and lacks ascomata, grow on rocks and trees along the central to southern CA coast; form *leucophaea* is found on the Channel Is. and farther south. *D. leucophaea* can grow larger than *D. alectoroides* and is more common; can be abundant in some locations.

A normal thallus of *Dendrographa leucophaea*. Photo shows form *minor,* from the Channel Is.

Dendrographa leucophaea parasitized by the fungus *Trimmatostroma dendrographae*. When taken over by this fungal parasite, the thallus assumes a different form and could be mistaken for an entirely separate species. Occasional, on coastal rocks in central and southern CA. Photo from Pt. Lobos, Monterey Co.

Evernia prunastri

Oakmoss lichen, Antler lichen, Staghorn lichen

Medium-sized fruticose lichen with flat branches in a dichotomously branched pattern. Distinct upper and lower surfaces, the upper being pale yellowish green and the lower surface white. Often with round soralia on the margins or surface of the branches, producing piles of gray or greenish granular soredia. Apothecia very rare. Resembles some species of *Ramalina,* but the white lower surface is diagnostic. Spot tests: cortex K–, KC+ gold, C–, P–; medulla, all tests negative. One of the most common and widespread lichens in CA, growing mostly on the bark of trees and shrubs, rarely on rock, statewide, except for the most arid parts of the Mojave Desert or high elevations. A rare form is white on both sides. In France, an extract of the lichen is used commercially as a fixative in perfume. Photo from the central Sierra Nevada foothills.

Kaernefeltia

Thornbush lichens

Small to medium-sized fruticose or foliose lichens, dark green or dark brown to almost black, often with spiny side branches and lecanorine apothecia but without soredia or isidia. Spores ellipsoid, colorless, 1-celled, 8 per ascus. Spot tests negative. On bark and wood. Can resemble *Tuckermannopsis,* but darker in color, and *Cornicularia,* which grows on rock and has apothecia at the lobe tips.

Kaernefeltia californica
Coastal thornbush lichen

Small, shrubby, fruticose, typically greenish black thallus with terete (rounded) branches up to 1.5 mm across, flattening at the axils, almost thorny in appearance. Apothecia tiny, lecanorine, along the branches with the branches folding under, giving the appearance of being at the tips. On twigs and wood, especially pine, in very coastal locations, statewide. Fairly common in the Pygmy Forest of Mendocino Co., but generally rather rare. Resembles *Nodobryoria abbreviata*, but blacker, not brown. Photo from the Pygmy Forest near Ft. Bragg.

K. californica

Kaernefeltia merrillii
Flattened thornbush lichen

Thallus small to medium-sized, foliose to somewhat fruticose, with flattened, dichotomously branched lobes, 0.3–3 mm wide, dark brownish green to almost black. The most common, more fruticose, narrow-lobed form is

This form of *Kaernefeltia merrillii* is quite fruticose. Photo from Mendocino Co. near Willits.

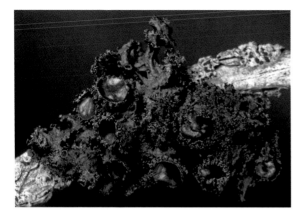

A fairly foliose form of *Kaernefeltia merrillii*. Photo the Rocky Mountains, MT.

widespread on bark and wood, especially on twigs of shrubs and trees, in coastal to inland habitats; the more foliose form with broadly flattened lobes grows at higher elevations on conifers. Can be abundant in some locations.

Leprocaulon americanum
Cottonthread lichen

Tiny fruticose lichen with minute, yellowish green stalks up to 6 mm long that have a granular surface, giving it the overall appearance of a granular crust, much like a species of *Lepraria*. Apothecia and perithecia absent. Spot tests: K– or weakly yellowish, C–, KC+ yellow, P– or weakly yellowish. On thin soil, moss, or plant debris, usually in shaded crevices in rock, in coastal locations statewide. Photo from the Channel Is. *Leprocaulon albicans* (not pictured), a rare species in CA, is P+ bright yellow, KC–.

L. americanum

Letharia

Wolf lichens

Medium-sized to large fruticose lichens, typically 3–15 cm long, with erect or pendant branches, ridged and rather stiff, 0.5–3 mm wide, very conspicuous because of the glowingly bright, lemon yellow to chartreuse color. Species can be quite variable in form. Spores colorless, 1-celled, 8 per ascus. Spot tests negative. Both species grow on the bark of conifers, notably red fir in the Sierra Nevada, or on wood, rarely on rock. In montane forests, the lower edge of *Letharia* growth on conifer trunks indicates the average height of the snowpack. In the Coast Range, *Letharia* is found often on wood, frequently on fences. On pines in eastern CA, *Letharia* often grows alongside *Vulpicida canadensis,* which is a similar color but is foliose and almost always has large apothecia. *Letharia* was used by Native Americans both as a poison and to produce a yellow dye.

Letharia columbiana

Brown-eyed wolf lichen

This is the fertile member of the two common species, and it is most often seen with dark brown apothecia. Branches with small, spiny branchlets; usually without soredia or isidia, but frequently with black pycnidia. Somewhat less common than *L. vulpina,* but it can be locally abundant in favored habitats, such as the upper montane zone on the western slope of the Sierra Nevada. *Letharia gracilis* (not pictured) is a recently named species that resembles *L. columbiana* but with branches that are smoother and more slender and drooping. It is known to date from the northern Klamath Range. Photo from the eastern side of the Cascades, OR.

L. columbiana

Letharia vulpina
Wolf lichen

The nonfertile species of Letharia. Apothecia very rare. Branches ridged and dotted with granular, coarse soredia that often become cylindrical isidia. Sometimes abundant on conifer bark and wood. Statewide, except for the Central Valley and Mojave Desert, mostly in the Sierra and interior ranges, but scattered in coastal areas as well. Photo from the Malheur National Forest, OR.

Niebla
Fog lichens

Fruticose lichens, pale green or yellowish green, with solid, round or flat branches. Medulla cottony or containing strands of supportive tissue. Apothecia common, lecanorine, sometimes pruinose. Branches often with black pycnidia. Spores 2-celled, 8 per ascus. Mostly on rock, but some species on shrubs. All are coastal, fog zone species found mostly in CA and Baja CA. They are similar to species of *Ramalina*, but they differ in having a much thicker and tougher cortex, and *Ramalina* lacks the conspicuous pycnidia found in most species of *Niebla*. Most species are quite variable in form.

N. cephalota

Niebla cephalota
Powdery fog lichen

Thallus small, often pendant, with very pale greenish branches up to 4 cm long, mostly round in cross section. Branches wrinkled and irregular, with distinctly bluish soralia appearing as powdery lumps. Apothecia absent. Medulla white, soft, and cottony. Spot tests: usually negative, but some populations react K+ yellow to orange or red, or P+ orange. On shrubs and trees, rarely on rock, along the entire length of the CA coast, occurring farther north than any other *Niebla* species. The only sorediate species of *Niebla*. Photo from Del Norte Co.

Niebla ceruchis
Bark fog lichen

Thallus small, bushy (caespitose), 2–5 cm long, quite variable, with rounded branches up to 4 mm wide, either pointed or blunt, lacking soredia; apothecia frequent, terminal or subterminal; branches often with pycnidia. Medulla white and cottony. Spot tests: usually negative, but occasionally K+ yellow to orange or red, or P+ orange. The only species of *Niebla* besides the smaller *N. cephalota* that commonly grows on shrubs and trees; occasionally found on rock. Mostly in southern CA, especially on the Channel Is.; a few specimens have been found as far north as Mendocino Co. Photo from the Channel Is.

Niebla ceruchoides
Clumping fog lichen
Thallus forms small, compact clumps of tightly bunched, roughly cylindrical branches up to 0.5 mm wide, branching in even dichotomies, resembling a small cushion plant or moss. Branches often with black pycnidia on the tips. Apothecia uncommon. Spot tests negative. Occasional on rock, sometimes on soil, San Luis Obispo Co. to Baja CA. Photo from the Channel Is.

Niebla combeoides
Bouquet fog lichen
Thallus forms clumps of fairly short cylindrical to angular, unbranched blades, up to 4 cm long, 0.5–2.5 mm width, with depressions and ridges. Apothecia

N. combeoides

abundant, usually on the tips, sometimes as many as 3 per tip. Pycnidia common, especially on the ridges. Branches are stiff when dry and will crack when bent. Medulla white, dense. Spot tests: usually negative, but occasionally K+ yellow to orange and P+ orange. On coastal rock in the fog zone, Marin Co. to Mexico. Compare with *N. procera* (longer blades, up to 8 cm, with some branching), *N. polymorpha* (blades crumpled and pitted, up to 6 mm wide when flat), and *N. robusta* (blades inflated and short, 2–4.5 mm wide, usually unbranched). Photo from Marin Co.

Niebla homalea
Armored fog lichen

Thallus forms clumps of flattened branches, mostly 1–5 mm wide but sometimes wider, a bit shiny, blackened at the base, with patterns of shallow depressions and ridges resembling plates, pointed at the tips or with apothecia. Medulla white, dense, with narrow, cordlike strands of tissue. One of the most common species of *Niebla*, and extremely variable in form. Spot tests negative. On coastal rocks, Marin Co. to Baja CA; on the Channel Is., some populations are loose on soil. Occurs farther inland than other *Niebla* species. Overall appearance is similar to *N. laevigata*, which has broader and smoother branches, typically 2–8 mm wide (sometimes up to 25 mm), without plates. The medulla of *N. laevigata* does not contain strands of hyphal tissue, and it is a bit greener with more frequent apothecia; there are chemical differences as well, although both have negative spot tests. Photo from Marin Co.

N. homalea

Niebla laevigata
Black-footed fog lichen

Thallus fruticose with broad (2–8 mm, but rarely up to 25 mm), shiny, rigid, flattened, and usually unbranched blades. Branches arise from a blackened base and crack if bent. Apothecia common, most often terminal, with whitish disks. Medulla thick, cottony, without strands of tissue. Spot tests negative. On coastal rocks facing the ocean, Monterey Co. to Baja CA. Compare with the more common *N. homalea*. Photo from the Channel Is.

Niebla polymorpha

Multiform fog lichen

Fruticose, dense, with round to flat, irregular, short, thick branches that divide only sparingly. Apothecia terminal, up to 5 mm wide. Medulla white, without hyphal strands. Spot tests negative, or K+ red and P+ yellow. On coastal rocks, San Luis Obispo Co. south, also farther inland than most species of *Niebla*. Compare with *N. combeoides*. Photo from the Channel Is.

N. polymorpha

Niebla procera

Tall fog lichen

Thallus forms dense clusters of more-or-less cylindrical, rigid branches up to 8 cm long, mostly 1–2 mm wide, blackened at the base and often on one side. Apothecia frequent, terminal or subterminal. Medulla white, without hyphal strands. Spot tests negative, or K+ red and P+ yellow. On rocks within the fog

N. procera

zone facing the ocean, San Luis Obispo Co. south. Compare with *N. combeoides*. Photo from the Channel Is.

Niebla robusta
Stumpy fog lichen

Thallus forms clumps of short, inflated, cylindrical, and mostly unbranched blades, 2–4.5 mm wide. Branches often with black pycnidia. Apothecia common, usually terminal. Medulla white, cottony, without hyphal strands. Spot tests negative, or K+ red and P+ yellow. On coastal rock, usually with *N. laevigata* and *N. procera*, San Francisco region to Baja CA. Has a stumpy appearance compared to other species. Photo from the Channel Is.

N. robusta

Nodobryoria
Foxtail lichen, Red horsehair lichen, Chestnut beard

Fruticose lichens rather like *Bryoria*, but they are distinctly reddish brown and have a unique, jigsaw puzzle–like cellular structure in the cortex. The slender, irregular, terete to flattened branches are pendant or shrubby and erect. Apothecia lecanorine; without soredia, isidia, or pseudocyphellae. Spores small, ellipsoid, colorless, 1-celled, 8 per ascus. Spot tests negative. Usually on conifer bark, occasionally on wood.

Nodobryoria abbreviata
Tufted foxtail lichen

Thallus shrubby, spiny-looking, red-brown, up 2.5 cm long. Apothecia common on or close to the branch tips; margins with spiny cilia. In size and form *N. abbreviata* resembles *Kaernefeltia californica*, but that species is very dark

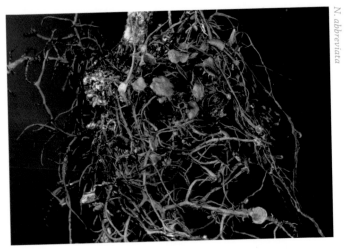

N. abbreviata

greenish gray and coastal. Primarily in drier, interior forests on pine and Douglas-fir, statewide except for the Mojave Desert. Photo from eastern WA.

Nodobryoria oregana
Pendant foxtail lichen

Thallus pendant, up to 17 cm long, with very slender, irregularly branched, dull reddish branches that are thicker at the base and with long furrows along the sides. Apothecia uncommon. *Bryoria fremontii* is somewhat similar, but that species usually has thicker main branches with perpendicular side branches, is more yellow-brown in color, and is usually shinier. Occasional on conifers, often at higher elevations, Coast Range and interior mountains

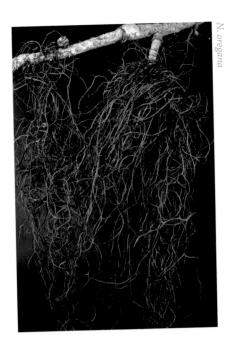

N. oregana

statewide, but uncommon in the south. Photo from Deschutes National Forest, OR.

Pilophorus acicularis
Devil's matchstick
Primary thallus is crustose, pale grayish green, often with large, pinkish to brown cephalodia. From this base emerge gray-green, verruculose stalks tipped with rounded, jet black apothecia. Mature stalks hollow, although other species of *Pilophorus* have solid stalks. Spot tests: K+ yellow, KC–, C–, P–. Occasional on rock in damp coastal habitats, northwestern CA. Photo of a damp specimen from Siskiyou Co.

P. acicularis

Polychidium muscicola
Moss thorns
Thallus small, subfruticose, forming a cushion of tiny, round, dark brown, shiny branches under 4 mm long and 0.2 mm wide that narrow toward the

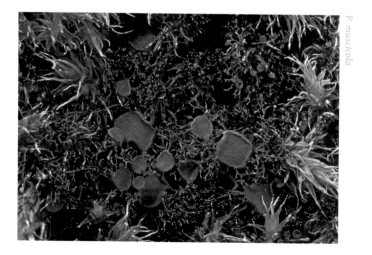

P. muscicola

tips. Apothecia common, biatorine, dark reddish brown. Photobiont is the cyanobacterium *Nostoc*. Spores ellipsoid to fusiform, colorless, sometimes pale red-brown, 2-celled, 8 per ascus. Spot tests negative. Occasional, on and with moss over rock or soil, statewide; more common in northern CA, mostly in mountain areas, often near streams. Compare with *Leptogium tenuissimum* and with *Pseudephebe*. Photo from Mt. Hood National Forest, OR.

Pseudephebe
Rockwool
Small, fruticose, dark brown to near-black thalli that usually form cushions resembling coarse hair, but sometimes form flattened branches and become almost foliose. Apothecia lecanorine; without soredia, isidia, or pseudocyphellae, but with abundant pycnidia appearing as tiny bumps with a hole on top. Spot tests negative. On rock in alpine locations. Can resemble *Polychidium muscicola*, but that species has a cyanobacterium photobiont, and *Pseudephebe* species grow in more exposed sites.

Pseudephebe minuscula
Coarse rockwool
Thallus small, almost black, usually with cylindrical branches 0.2–0.5 mm thick, narrowing at the tips and closely appressed to the rock. Often with flatter, almost foliose branches up to 1 mm wide at the margins, or becoming almost crustose and irregularly areolate in central parts of the thallus. Branching irregular, with short distances between the nodes. Apothecia common. The photo shows the *Melanelia*-like morph; in its other form, it has a mat of slender branches and can resemble *P. pubescens* but has shorter internodes. Specimens with flat branches can also resemble *Cornicularia normoerica*, but that species has thicker, more sparingly divided, more erect branches and terminal apothecia. On exposed rock in the alpine Sierra Nevada, northern ranges, and southern inland mountains. Photo from Southeast AK.

P. minuscula

Pseudephebe pubescens

Fine rockwool

Thallus forming clumps of dark brown to black, fairly cylindrical, fine, forked branches less than 0.2 mm wide, often interwoven. Apothecia uncommon. Can resemble the more slender forms of *P. minuscula*, but the branches are longer, finer, and rounder, and the distance between internodes is longer. On siliceous rock, very rarely on hard wood, in alpine habitats, with a range similar to that of *P. minuscula*. Photo from Southeast AK.

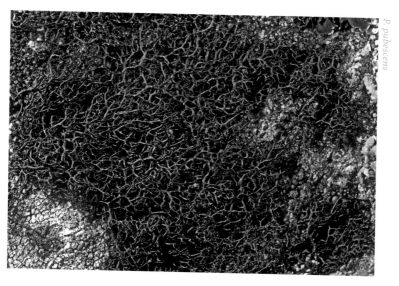

P. pubescens

Ramalina

Ramalina, Strap lichens

Small to very large fruticose lichens, pendant or shrubby, with flattened, straplike, greenish, rather stiff branches. Some species are sorediate. Apothecia lecanorine. Pycnidia infrequent, pale. Spores ellipsoid, colorless, 2-celled, 8 per ascus. Spot tests: cortex K–, KC+ dark yellow, C–, P–, medullary reactions various. Mostly on bark, but a few species prefer rock; *Ramalina* species are among the more common and conspicuous lichens in the state, often growing alongside species of *Usnea*. Compare with species of *Niebla*, which are usually dotted with black pycnidia; *Usnea*, which have more rounded branches and a central cord; and *Evernia prunastri*, which is almost foliose with a white lower surface.

Ramalina canariensis

Canary Ramalina

Thallus small with short, rather broad lobes (up to 5 mm), forming small cushions or shrubby masses. Lobes split at the edges or form irregular holes, exposing masses of fine soredia; lobes can develop longitudinal ridges or wrinkles. Spot tests: medulla K–, KC–, C–, P–. On shrubs and trees near the coast, San Francisco area to Mexico. Photo from San Francisco.

Ramalina farinacea

Dotted Ramalina

Thallus small to medium-sized, quite variable but usually with rather narrow, straplike branches 3–7 cm long, 0.5–3 mm wide, most often somewhat pendant. Branches with discrete, round to elongated soralia, mostly along the margins but often on the surface as well, with pale green, powdery (farinose) soredia. Apothecia uncommon, disks up to 6 mm, pruinose. Spot tests: medulla and soredia negative, or sometimes K–, KC+ pink, C–, P+ red-orange, or K+ red, KC–, C–, P+ yellow. Usually on bark, rarely on rock. Although it

does not form large, showy masses like *R. menziesii, R. farinacea* is perhaps as common, occurring along the entire length of the Coast Range region, less often in the Sierra Nevada and other inland mountains. The less common species, *R. subleptocarpha,* is also sorediate, but the soredia form continuous lines along the lobe margins. Photo from Dosewallips State Park, WA.

Ramalina leptocarpha
Western strap lichen
Thallus small to medium-sized, erect or pendant, with rather broad branches 2–6 mm wide, 3–15 cm long, with shallow longitudinal depressions and ridges. Without soredia. Apothecia flat or concave, common on the margins, sometimes on the surface. Spot tests negative in the medulla. Fairly common on bark, especially of broad-leaved trees, generally coastal or semicoastal statewide, more common in central and southern CA. This is the fertile counterpart of the sorediate species *R. subleptocarpha.* A similar, rarer species from the central and southern Coast Range is *R. puberulenta* (not pictured), which has cortical hairs on the branches. Photo from Monterey Co.

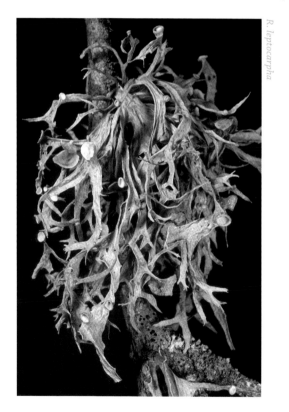
R. leptocarpha

Ramalina menziesii
Lace lichen, Fishnet lichen
Thallus pendant, up to 1 m long, with branches that vary from narrow to broad (up to 3 cm) often on the same thallus, with raised ridges and striations, and

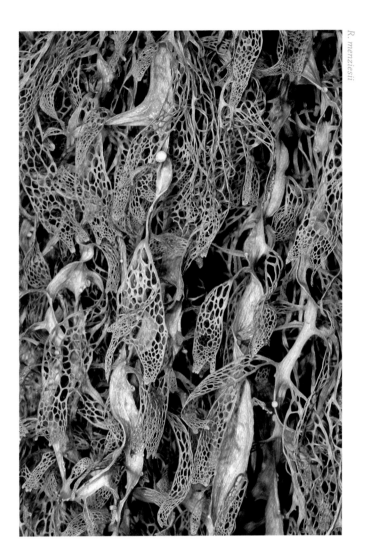

R. menziesii

side branches or tips that form distinctive nets unlike any other North American lichen. Color varies, usually pale yellowish green, deeper green when wet. Without soredia. Apothecia fairly common but not always present. Spot tests negative in the medulla. On bark, especially of oaks, but sometimes on conifers, often but not exclusively in riparian habitats. In favored locations it can drape trees with spectacular masses of hanging green strands. Statewide, frequent in the Coast Range, but known also from a few locations on the western slope of the Sierra Nevada. Extremely variable in growth form: in some coastal habitats, the nets can be tiny and inconspicuous, occurring only on some branch tips, with branches so slender that the thallus resembles a species of *Usnea;* and in dry, inland locations, it can become short, shrubby, and almost thorny-looking. An important food for deer in parts of the Coast Range; they will rear up on their hind legs to get to the higher strands. A number of bird species use it for nesting material. Photo from San Luis Obispo Co.

Ramalina pollinaria
Chalky Ramalina

Thallus small, shrubby, up to 2.5 cm long, pendant or erect, with variable branches that have powdery, granular soredia near the tips or mostly on the undersurface of torn-open lobe tips. Spot tests negative in the medulla. Not especially common on bark or rocks in the shade, almost always in the Coast Range, but occasionally in the northern Sierra Nevada or southern mountains. Can resemble *R. farinacea,* but that species has soralia all along the branches, not concentrated at the tips. Photo from Humboldt Co.

R. pollinaria

Ramalina roesleri
Frayed Ramalina

Thallus small, shrubby, 1–4 cm long, with very narrow, almost round branches less than 1 mm wide, arising from somewhat broader and flatter main stems that are hollow and have small perforations. The lobe tips curl and break up into granules and coarse soredia. Apothecia absent. Spot tests negative in the

R. roesleri

medulla. Uncommon on twigs and branches in shady, humid locations, rare on rock, far northwest coastal region only. Photo from southern coastal OR.

Ramalina subleptocarpha
Slit-rimmed Ramalina

Thallus small to medium-sized, pale yellowish green, usually pendant, 3–10 cm long, with fairly narrow branches, typically 2–4 mm wide, that have lines of soredia all along their edges, rarely also developing on the surface. Spot tests negative in the medulla. *R. subleptocarpha* is the sorediate counterpart of *R. leptocarpha*. The long lines of marginal soredia are distinctive. On bark and wood in open areas, often on coastal shrubs, scattered in coastal areas statewide. Photo from coastal San Luis Obispo Co.

R. subleptocarpha

Roccella
Orchil lichens, Canary weed

Light gray, medium-sized, fruticose lichens, usually pendant, with flattened to cylindrical branches. Fruiting bodies resemble lecanorine apothecia. Spores usually fusiform, straight or curved, colorless, 4-celled. Rather rare along the central and southern coast, more common along the west coast of Baja CA. Some species were heavily collected in earlier times to produce orchil, a purple dye; a European species of *Roccella* is used to produce litmus. Compare with *Dendrographa*, a more common genus sharing a similar habitat.

Roccella decipiens

Broad-strap orchil lichen

Thallus pendant, with gray, or slightly brownish gray, flattened branches 5–20 cm long, usually wrinkled and irregular. Apothecia common, circular, with wavy margins; without soredia. Spot tests: cortex K+ yellowish red (sometimes faint), KC+ red, C+ red, P–; medulla no reactions. Has been called *R. fimbriata*. Rare on rock, wood, or bark, strictly coastal, San Francisco area to Mexico. Photo from Baja CA.

Roccella gracilis

Slender orchil lichen

Thallus pendant, pale gray, with branches that are usually flattened near the base and for much of the thallus, 1–4 mm wide, 4–12 cm long, becoming slender and more cylindrical near the tips, with ridges and wrinkles and abundant farinose soredia. This is probably the sorediate counterpart to *R. decipiens*, which almost always produces apothecia, and it has the same spot test reactions. Rare on bark and vertical rock, only in southern coastal sites. This species has been called *R. babingtonii* and, later, *R. peruensis*, in the North American lichen literature. Photo from Baja CA.

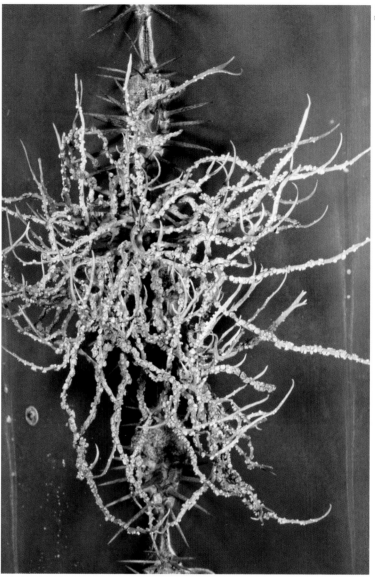

Schizopelte

A small genus of fruticose to crustose lichens with creamy white to tan thalli and somewhat irregular, often partially immersed apothecia. All species are strictly coastal.

Schizopelte californica
Fog fingers

Thalli small to medium-sized, fruticose, forming cushions, usually with a mix of rounded and flattened, erect, thick, tan to creamy gray stalks 1–3 mm wide, 10–30 mm long. Some branches typically broaden into apothecia with lecanorine margins, violet-gray in color, and pruinose, variable in size and shape. Spores brown, 4–8 celled, 8 per ascus. Spot tests: cortex K–, KC+ red, C+ red to yellow; medulla no reactions. A distinctive, strictly maritime species, San Francisco Bay to Baja California. Generally rare but can be locally abundant, especially on the Channel Is. Photo from the Channel Is.

S. californica

Schizopelte parishii
Brittle bag lichen

Thallus small (1–2 cm high), fruticose but lobate, creamy pale gray or yellowish with hollow, inflated, quite brittle lobes that produce conspicuous clusters of granular soredia on their tips. Fruiting bodies (ascomata) are rare, scriptlike. Spores fusiform, colorless, 4–5 celled, 8 per ascus. Spot tests: cortex K– or K+ yellowish, KC+ red, C+ red, P–; medulla K–, KC+ red, C+ red, P–. On vertical rock near the ocean, central to southern coast. Endemic to this region and southward into Baja CA. Photo from the Channel Is.

S. parishii

Seirophora californica

Gray bush lichen

Thallus fruticose, erect, pale gray, rarely over 3 cm long, with flattened, wrinkled branches 0.2–1.3 mm wide, broad at the base and then tapering and much divided, usually with a basal holdfast. Often with farinose, greenish gray soredia, especially near the lobe tips. Apothecia uncommon, orange, often with protruding, yellow to orange, wartlike pycnidia; spores polarilocular. Spot tests negative, except apothecia and pycnidia react K+ purple. Thallus often has a fungal parasite, *Sphaerellothecium subtile*, forming a black network on the surface. Formerly in the genus *Teloschistes*. Found almost exclusively on twigs of trees and shrubs in the Channel Is. Photo from the Channel Is.

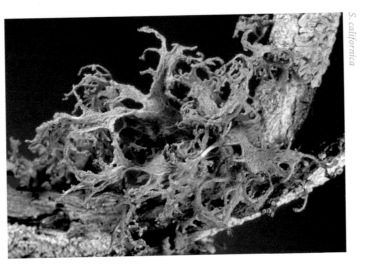

S. californica

Sphaerophorus

Coral lichens, Tree coral

Shrubby fruticose lichens with slender, cylindrical, solid, much-divided branches whose tips often have spherical apothecia that are pale on the outside but burst open with a mazaedium inside, revealing a mass of black spores. Spores ellipsoid, brown, 1-celled. Most often on bark in northern forests, sometimes on rock. Both species below were formerly lumped together as *S. globosus*, an arctic-alpine species on soil and rocks that, in the strict sense, does not occur in California.

Sphaerophorus tuckermanii

Tuckermann's coral lichen

Thallus fruticose, near-white or greenish to pale grayish brown, with slender, smooth branches bearing dense clusters of brittle coralloid branchlets. Apothecia, if present, fairly sparse, globose. Compare with *S. venerabilis*, below. On bark, especially of conifers, mostly in old-growth Coast Range forests, Santa Cruz Co. to OR. Photo from the Gifford Pinchot National Forest, WA.

Sphaerophorus venerabilis
Ancient coral lichen

Thallus fruticose, with branches that usually have many dents and depressions; without clusters of coralloid branches, or very sparse if present. The branches can sometimes have a somewhat dorsiventral differentiation if growing out from a tree trunk in shelflike clumps, with the upper side redbrown to grayish and the lower side pale gray. Apothecia globose, common at the branch tips. Similar habitat and range as *S. tuckermanii,* and the two species often grow together, but *S. venerabilis* favors a drier, less coastal habitat. Photo from Six Rivers National Forest.

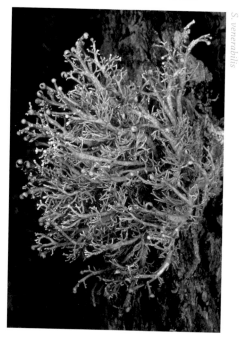

Stereocaulon sasakii
Woolly foam lichen

Thallus fruticose, white, shrubby, either forming a prostrate mat or growing erect. A central column supports phyllocladia, flattened squamules along the stalks that contain the green alga *Trebouxia*. Stalks are usually covered in a creamy tomentum in which are buried granular, dark blue-green cephalodia. Spot tests: P– or pale yellow. Occasional on soil or mossy rock, northern mountains. Morphologically almost identical to *S. tomentosum* (not pictured), a common species of boreal North America, differing mostly in containing lobaric acid. *Stereocaulon* is a widespread genus with many species in arctic and alpine regions but is rare in CA: *S. glareosum* and *S. rivulorum* have been found in the Sierra Nevada, and there is one record of *S. sterile* from the central coast; *S. intermedium* probably also occurs in the state. Photo from Mt. Hood National Forest, OR.

S. sasakii

Sulcaria badia
Bay horsehair lichen

Thallus fruticose, pendant, 20–50 cm long, rather like a species of *Bryoria* but with a distinctive pale grayish tan to reddish brown color. Branches mostly cylindrical, but main branches are often flattened and deeply grooved with long spirals of pseudocyphellae. Spot tests: thallus K+ yellow, KC± yellow to pink, C–, P+ pale yellow to brownish. Rare, found only in a few sites in the

Coast Range in the San Luis Obispo area and Mendocino Co. A related species, but even rarer, is *Sulcaria isidiifera* (not pictured), similar in color to *S. badia*, with long soralia that split open to reveal isidia, found in old-growth coastal scrub. Photo from near Laytonville, Mendocino Co.

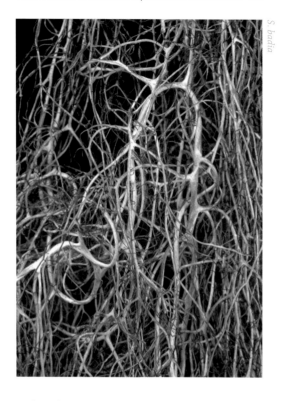

S. badia

Teloschistes
Orange bush lichens
Small to medium-sized, bright orange, fruticose lichens, mostly on twigs, but occasionally on rock or soil. Spores polarilocular, 8 per ascus. Spot tests: orange parts of cortex and apothecial disks K+ deep red-purple, KC–, C–, P–. The genus is strikingly different from any other, although *Seirophora*, which has a gray thallus, is related.

Teloschistes chrysophthalmus
Gold-eye lichen
Thallus small, shrubby, with short, flat to reticulately ridged main branches, pale to bright orange above but whitish on lower surface, with tips that often have short cilia. Apothecia almost always present, large, orange, deeper in color than the branches. Occasional on branches and twigs in the Coast Range and southern coastal habitats, Mexico to Sonoma Co., becoming more common in southern CA. Photo from Baja CA.

Teloschistes flavicans
Powdered orange bush lichen

Thallus small to medium-sized, shrubby, bright orange, with cushions of slender, rounded to angular branches that are corticate on all sides. Branches often bear abundant yellowish soralia that break through the cortex, or the thalli can be without soralia, producing instead abundant apothecia near the

The sterile, sorediate form of *Teloschistes flavicans*. Photo from the Channel Is.

The fertile form of *Teloschistes flavicans*; previously called *T. exilis*. Photo from Sonoma Co.

branch tips. Apothecia are not ciliate, as in *T. chrysophthalmus*. Occasional on twigs, sometimes on the ground, in coastal areas statewide, but most often in southern CA. Earlier records of *T. exilis* from CA are now thought to be mostly or entirely fertile specimens of *T. flavicans*. *T. exilis* has broader lobes, cilia that are not reddened, and different chemistry; if it occurs at all in CA, it is rare.

Usnea

Beard lichens, Old man's beard

Fruticose lichens that can be small and shrubby, or pendant, sometimes very long; most species are yellowish green, but two in CA are reddish. Branches round or angular, with a central cord of supporting tissue, the axis. The cortex is often fissured, creating segments along the branches; the surface may be smooth or have papillae (tiny, translucent bumps made up of cortex tissue), tubercules (warts containing medullary tissue), or fibrils (short, perpendicular side branches that give the branches a spiny look). Soredia and isidia are frequent, often with both on the same thallus. Apothecia are lecanorine. The branching can have distinct main stems with side branches forming an asymmetrical pattern, or it can be symmetrical like the letter "Y," referred to as dichotomous, with the stems of the fork equal or unequal in size. The relative thicknesses of the cortex, medulla, and central axis are important in distinguishing species. Spores small, ellipsoid, colorless, 8 per ascus. Spot tests: cortex KC+ dark yellow in all species (usnic acid); medulla with varied reactions. *Usnea* is a widespread and common genus with many species, and the amount of variation often makes naming them challenging; chemical tests are particularly important. Larger pendant species are sometimes confused with *Alectoria sarmentosa,* but no *Alectoria* has a central cord. Usnic acid is a topical antibiotic that has often been used in a variety of medications, but many people are allergic to it. *Usnea* species grow mostly on bark and twigs, occasionally on wood; a few species can grow on rock. They can be abundant and conspicuous on the upper branches of trees in Coast Range woodlands or on shrubs in chaparral.

Usnea cavernosa

Pitted beard lichen

Thallus long, pendant, 10–60 cm long, dichotomously branched. Main branches constricted at the nodes, with conspicuous dents and ridges. Without soredia, isidia, or papillae. Apothecia very rare. Cortex thin, medulla and axis broad. Spot tests: medulla K+ yellow turning red or K–, KC–, C–, P+ yellow or P–. Rather uncommon, mostly in conifer forests, Coast Range. Used by the Wylackie people in northwestern CA to tan leather. Photo from southern ON.

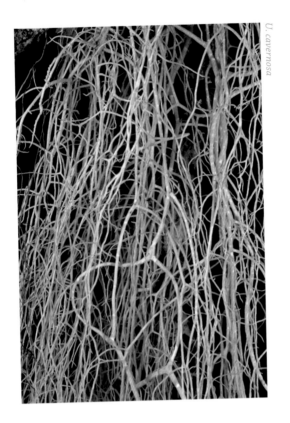

U. cavernosa

Usnea ceratina
Warty beard lichen

Thallus coarse, shrubby to pendant, up to 30 cm long, and scraggly, with warty tubercules on the larger branches that turn white at the tips, making the thallus white-spotted. Branching is more or less equally dichotomous. Papillae on the branches, and often with isidia, soralia-like isidia, or pointed fibrils. Apothecia absent. Cortex thick and tough. Medulla reddish (rarely white), thick. Axis pinkish, broad. Spot tests: all negative or sometimes K+ yellowish, KC yellowish, C+ yellow. Occasional on conifers and shrubs statewide, mostly in the Coast Range. Compare with *U. mutabilis*. Photo from Del Norte Co.

Usnea cornuta
Small inflated beard lichen

Thallus small, shrubby, becoming pale or somewhat blackened at the base. Branching dichotomous but with stems of unequal size, with branches 2–5 cm long, narrow at the base, appearing inflated, often with numerous fibrils. Papillae on the surface, and usually with soredia, especially near the tips, with isidia mixed in. Apothecia very rare. Cortex thin, shiny. Medulla broad, sparse, and cobwebby ("lax"). Axis variable. Spot tests: medulla and axis usually K+ yellow to red, KC–, C–, P+ orange; sometimes K+ yellow, KC+ pink, C–, P+ orange; rarely K– or yellow, KC–, C–, P+ yellow. In Coast Range forests

statewide. A subspecies, *U. cornuta* subsp. *cornuta*, occurs on the central coast but is rare. Resembles *U. subfloridana*, but that species lacks basal constrictions and has a blackened base and dense medulla. A similar species is *U. dasaea* (not pictured), which has dense fibrils on parts of its branches, giving it a spiny look; it occurs on the southern coast. Photo from Roan Mountain, TN.

Usnea esperantiana
Hopeful beard lichen

Thallus, small, shrubby, up to 5 cm long. Main branches cylindrical to irregular with lateral branches that are twisted and recurved, constricted at points of attachment, giving the branches an "inflated" appearance. Branching pattern varies from fairly dichotomous to irregular, with main stems and smaller side branches; branches divergent, segmented. With small papillae and short, sparse fibrils, but with broad soralia that are irregular in shape, often almost covering the tips of branches and fibrils. Without isidia. Apothecia absent. Cortex thin, shiny. Medulla webby. Axis thin, white to reddish. Spot tests: medulla K+ yellow turning red, KC–, C–, P+ orange to yellow. Occasional on bark, especially of oaks, sometimes on wood, rarely on rock, in woodlands and chaparral, central and southern Coast Range. Photo from near Cambria, San Luis Obispo Co.

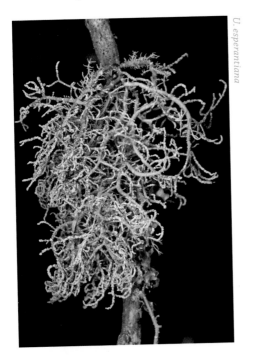

U. esperantiana

Usnea filipendula
Fishbone beard lichen

Thallus pendant, often over 20 cm long. Several main branches typically darker than side branches; many side branches and fibrils, main branches blackened at the base. Branches with abundant fairly long, cylindrical papillae, also with isidia in clusters, or single. Apothecia absent. Cortex fairly thick. Medulla dense and thin. Axis somewhat thick. Spot tests: medulla K+ red, KC–, C–, P+ yellow. On bark, usually of conifers, mostly in the Coast Range, but also on the western slope of the Sierra. Compare with *U. scabrata*, which is paler, has a thinner, brittle cortex, and differs in chemistry. Photo from southern ON.

Usnea flavocardia
Red-spot beard lichen

Thallus shrubby, fairly small and erect, less than 7 cm long, with some branches pale green and others whitish but red-spotted. Base either like branches or brown to almost black. Axis color almost unique in *Usnea* in being yellow, at least in part. Branches cylindrical, mostly with main stems and smaller, divergent side branches, but sometimes dichotomous, usually with numerous soralia and cylindrical papillae. Apothecia absent. Cortex thin. Medulla variable but usually thick and in two layers with the outer one dense, but lax near the axis. Axis variable in width. Spot tests: medulla usually K+,

KC+ and C+ all bright yellow, P+ orange; sometimes all tests negative. Fairly common on bark, sometimes on rock or wood, along entire length of coast. Photo from coastal Humboldt Co.

Usnea fragilescens
Long inflated beard lichen
Thallus shrubby, with divergent branches 3–8 cm long and a broad, blackened base. Branches constricted at attachment, appearing inflated; branching unequally dichotomous. May be smooth or with abundant papillae, usually with perpendicular fibrils, more or less sorediate, with large, circular soralia scattered over the branches that can be excavate or mounded, often with isidia. Apothecia absent. Cortex thin and shiny. Medulla very broad, lax. Axis narrow. Spot tests: medulla K+ yellow to red, KC−, C−, P+ yellow. On trees and shrubs along entire length of coast. Compare with *U. glabrata*, which differs in chemistry and never has isidia. Photo from coastal Humboldt Co.

U. fragilescens

Usnea glabrata
Lustrous beard lichen
Thallus shrubby and erect, with short tufts that are black at the base. Branching unequally dichotomous; branches swollen, constricted at the base, with abundant excavate to tuberculate soralia concentrated close to the tips, but infrequent papillae. Without isidia, sometimes with fibrils. Apothecia absent. Cortex thin. Medulla broad and lax. Axis narrow. Spot tests: medulla K−, KC− or pink, C−, P+ red-orange. On bark along entire length of coast. Resembles *U. flavocardia* in structure but lacks the reddish spots. Photo from Monterey Co.

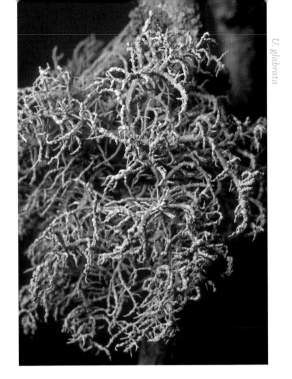

Usnea hirta

Bristly beard lichen, Shaggy beard lichen

Thallus shrubby, short, up to 5 cm, yellow-green and not blackened at the base. Branches with ridges, angular in cross section, often with fibrils but

without papillae; branching is unequally dichotomous. Abundant isidia along branches, but without soredia. Cortex thin. Medulla thick, white, lax. Axis narrow. Spot tests: medulla negative except rarely K+ yellow becoming red. Often on hard, dead wood, or on bark, especially of conifers, or sometimes on rock, statewide, coast and inland. Photo from Modoc National Forest.

Usnea intermedia
Western bushy beard

Thallus shrubby, stiff, 3–12 cm across, sometimes somewhat pendant; branching is unequally dichotomous with numerous short, irregular branches and fibrils protruding perpendicularly from the main branches; with papillae, especially on older parts of the thallus. Apothecia abundant, round, pruinose, paler yellow-green than the branches, with spiny margins. Cortex thick, tough. Medulla moderately broad and lax. Axis somewhat broad, usually about the same width as the medulla. Spot tests: medulla K+ yellow turning red, KC–, C–, P+ orange-yellow or P–; or all reactions negative. Fairly common on trees, especially oaks, rarely on wood or rock, in the central and northern Coast Range. This is the species formerly called *U. arizonica* in North America. Photo from Marin Co.

Usnea lapponica
Powdered beard lichen

Thallus of shrubby tufts up to 12 cm long, sometimes subpendant, generally with a few main branches but many perpendicular side branches that do not narrow at their attachment points; with numerous fibrils. Branches with abundant papillae and deeply excavate, concave soralia with powdery soredia. Without isidia. Apothecia absent. Cortex thin. Medulla variable but usually thick, white, cottony, or dense. Axis also variable. Spot tests: medulla

U. intermedia

negative, or sometimes K+ yellow or red, KC–, C–, P+ yellow. On conifers, mostly in the Coast Range, but also on the western slope of the Sierra Nevada. Resembles a rarer species from central and southern coastal areas, *U. fulvoreagens* (not pictured), which has a dichotomous and equal branching pattern, tapered branches, and thinner medulla; it differs in chemistry. Photo from Rocky Mountain National Park, CO.

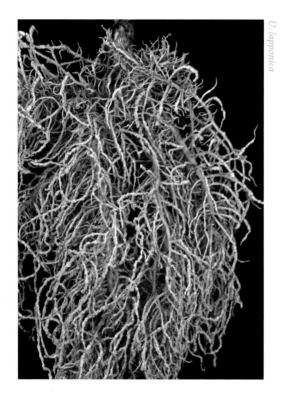

U. lapponica

Usnea longissima
Methuselah's beard

Thallus pendant, up to 3 m long, the longest lichen in the world. Main branches have few divisions but numerous short side branches and fibrils of about equal length. Side branches occasionally bear eroding cortical areas resembling soredia. Cortex smooth but disintegrates in patches on main stems, letting the medulla show through over a pink-brown central cord. Spot tests negative in the medulla. An uncommon and beautiful forest species, this remarkable lichen grows in scattered locations in the northern Coast Range, often along a river or stream. It is always a special treat to see its thin strands hanging in graceful garlands from a group of trees. It was the original Christmas tree tinsel, used for that purpose in Scandinavia, but now the species is rare in Europe, probably owing to increasing air pollution, as it is highly sensitive to air quality. Photo from western slope of the Cascades, OR.

Usnea mutabilis
Bloody beard lichen
Thallus shrubby to somewhat pendant, 3–7 cm long, often a bit darker green than most *Usnea* species. Branching is unequally dichotomous, with only a few perpendicular fibrils. Surface isidiate, with abundant isidia in clusters or isolated; can also be papillate or warty. Apothecia absent. Cortex variable in width, brittle. Medulla dull red, thin, dense. Axis thick, also dull red. Spot tests negative in the medulla. Mostly a species of eastern North America, occasional in southern CA, mostly on chaparral shrubs or pines, rarely on rock. *U. ceratina* also has a reddish medulla but is more pendant and has whitish tubercules. Photo from San Luis Obispo Co.

Usnea rubicunda
Red beard lichen
Thallus shrubby to somewhat pendant, up to about 15 cm long, with varying amounts of red or orange color. Branching is unequally dichotomous, di-

vergent; branches usually have abundant isidiate soralia that appear as white spots and are often papillate as well. Base is pale, not blackened. Cortex thick, glassy. Medulla thin, white, dense. Axis broad, colorless. Spot tests: medulla usually K+ yellow, KC–, C–, P+ orange; sometimes K+ red, P+ yellow. A distinctive species, and fairly common on shrubs and trees, especially pines, in

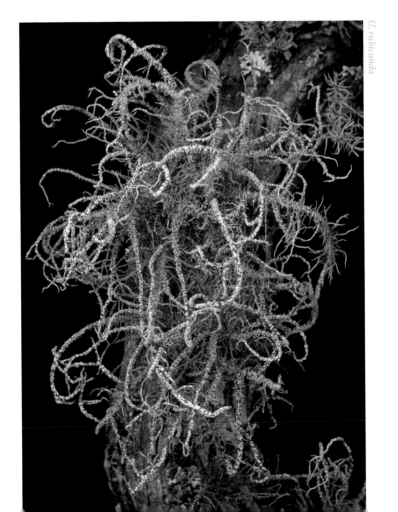

open coastal forests along entire length of coast. From a distance very orange specimens might be mistaken for *Teloschistes flavicans,* but most thalli of *U. rubicunda* have some pale green or whitish portions, and *Teloschistes* species are not white-spotted. Photo from San Luis Obispo Co.

Usnea scabrata

Straw beard lichen

Thallus very pale, yellowish green, with slender, pendant branches 5–15 cm long, branched at the base, usually with abundant isidia and papillae, and often with short fibrils and warts. Branching is unequally dichotomous; branches irregular, with ridges and depressions, and tips that branch and taper. Cortex thin. Medulla white, lax. Axis slender. Spot tests sometimes show no reactions in the medulla, but otherwise K+ yellow turning red, KC–, C–, P+ yellow-orange. Occasional on bark, mostly in the Coast Range, but also in northern interior mountains. Can resemble *U. filipendula,* but that species has darkened main branches near the base, a thick cortex, and a thin, dense medulla, and it contains salazinic acid. Photo from a herbarium specimen, collected in BC.

U. scabrata

Usnea subfloridana
Delicate beard lichen

Thallus shrubby and erect, with rather slender, tapering branches up to 7 cm long and a distinctly blackened base. Branching is dichotomous, divergent, dense, usually with numerous well-delimitated soralia containing short isidia, often with papillae. Apothecia absent. Cortex thick. Medulla usually thin, dense. Axis moderately thick, unpigmented. Spot tests: medulla usually K+ intense yellow turning orange, KC–, C–, P+ orange; sometimes all medulla reactions negative. Occasional on bark, especially oaks, mostly in the Coast Range but also northern Sierra. Compare with *U. cornuta* and *U. fragilescens*. Another similar, but rarer, species is *U. praetervisa* (not pictured), with more diffuse soralia and different chemistry; it occurs on the southern coast. Yet another species that resembles *U. subfloridana* is *U. diplotypus* (not pictured), which has dichotomous but unequal-sized branching, sometimes lacks a blackened base, and reacts medulla K+ red, C–, KC–, P+ yellow; its frequency in California is uncertain. Photo from Mt. Tamalpais, Marin Co.

U. subfloridana

Also of Note

A number of other *Usnea* species, not described here, occur in CA. Most are quite uncommon, but one that is seen fairly frequently in central and southern CA is *U. brattiae* (not pictured), which has a shrubby or slightly pendulous thallus with more or less dichotomous branching, cylindrical to tapered branches that don't narrow where they attach, and punctiform soralia with isidiomorphs; found mostly in coastal regions.

Crustose and Squamulose Lichens

- -

Acarospora

Cobblestone lichens, Cracked lichens

Crustose lichens with an areolate thallus, sometimes almost squamulose. Some species are brilliant yellow, others can be white, brown, or gray. Apothecia usually immersed, round to irregular. Spores ellipsoid to spherical, colorless, dozens to hundreds per ascus. Usually on rock but a few species on soil. Brilliant yellow species of *Acarospora* resemble *Pleopsidium flavum,* which differs in the structure of the ascus and cortex tissue; nonyellow ones can resemble species of *Aspicilia.*

Acarospora elevata

Mountain cobblestone lichen

Thallus, dark reddish brown, areolate or verrucose, usually glossy. Apothecia large, rough, black (sometimes dark red), sunken, epruinose. Spot tests negative, UV–. On exposed granitic rock, central and southern CA mountains.

A. elevata

Most often found at higher elevations but occasionally coastal; often eaten by red mites. Photo from Joshua Tree National Park.

Acarospora fuscata
Brown cobblestone lichen

Thallus brown, areolate, continuous or dispersed, areoles often shiny and sometimes lifting at the edges. Apothecia dark brown, irregular. Spot tests: cortex (sometimes medulla) K–, KC+ red, C+ pink but hard to see, P–. Occasional on rock, especially granite, central to southern CA and eastside of Sierra. A similar species, common in the Sierra and southern mountains, is *A. thamnina* (not pictured); it is shiny, often cracked like *A. fuscata,* but is truly squamulose with some overlapping scales and reacts strongly C+ red. Photo from Mt. San Jacinto, Riverside Co.

A. fuscata

Acarospora obnubila
Cloudy cobblestone lichen

Thallus squamulose, with dull brown, convex squamules up to 2 mm wide, round or irregular, with a stipe sometimes obscured by soil particles; can appear almost areolate. Apothecia small, reddish brown, 1–6 per squamule, often with abundant pycnidia. Frequently appears as scattered squamules growing among other lichens; stipe development is variable. Spot tests negative, UV–. On rock, southern CA mountains. Photo from Joshua Tree National Park.

Acarospora schleicheri

Soil paint lichen

Thallus pale to bright yellow, sometimes greenish, especially if damp, not shiny, areolate to squamulose. Apothecia dark brown, embedded, rounded, 1–3 per areole. Can be lobed at the margins, and sometimes pruinose. Spot tests negative, UV+ orange. Fairly common on soil in drier habitats, central to southern CA. Photo from the Channel Is.

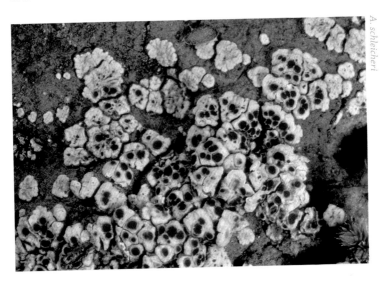

Acarospora socialis

Bright cobblestone lichen

Thallus brilliant yellow or yellow-green, usually areolate with irregularly shaped areoles but often becoming squamulose; areoles sometimes scattered,

An inland form of *Acarospora socialis*. Photo from Tonto National Forest, AZ.

or following cracks in the rock, or forming patches up to 10 cm across; sometimes pruinose. Apothecia usually tan or brown, rarely yellow or black, immersed, most often one per squamule but sometimes more in arid habitats. An extremely variable and common species, often called by several different names in the past. Coastal forms are often greener in color than desert ones. Sometimes found with any of several fungal parasites. Spot tests negative, UV+ orange. Found on rock, rarely on soil, coast and inland statewide, but more common toward central and southern CA. Can resemble *Pleopsidium flavum*, but that species usually has areoles with lobed margins and smaller, yellower apothecia less than 1 mm wide, and the ascus walls and tips turn dark blue with iodine (K/I); it is usually found at elevations over 900 m. *A. socialis* also resembles *A. robiniae* (not pictured), which has a C+ red cortex and more closely attached areoles; it is a maritime species found from San Luis Obispo Co. to Mexico.

A greener, coastal form of *Acarospora socialis*. Photo from coastal Sonoma Co.

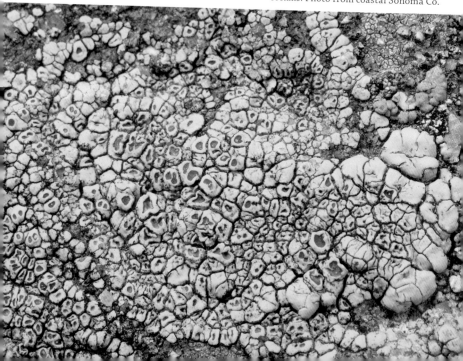

Acarospora strigata
Hoary cobblestone lichen

Thallus areolate to verruculose, brown, but often whitish or pale gray to yellowish when covered with pruina; areoles up to 3 mm wide but can be tiny, less than 0.5 mm. Apothecia dark red-brown, one to several per areole, sometimes filling it; disks sometimes pruinose. Areoles often cracked, especially around the apothecia. Spores broadly ellipsoid. Spot tests negative, UV–, but sometimes a pale color due to the presence of minerals. Mostly in southern CA deserts and mountains. Photo, from the Mojave Desert, shows *A. strigata* "group."

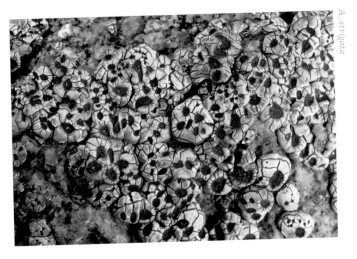

A. strigata

Acarospora thelococcoides
Soil eyes lichen

Thallus verruculose, whitish to brown, in patches up to 10 cm across and usually pruinose, forming white rings around the immersed, dark brown apothecia. Spot tests negative, UV–. On soil, especially granitic, southern CA coastal

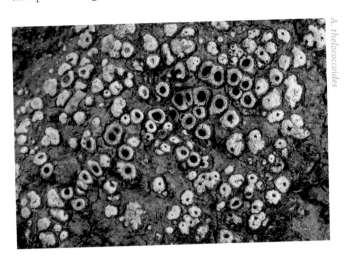

A. thelococcoides

mountains. An indicator of undisturbed soils, often occurring with the lichen *Texosporium sancti-jacobi.* Photo from the Coast Range near Santa Barbara.

Also of Note

Quite a few other species not described above occur in CA. Three of the more common ones are:

Acarospora americana, with a medium to dark brown or black thallus, areolate to verruculose or squamulose, without pruina or heavily pruinose; apothecia deeply immersed, reddish to blackish brown; spot tests negative; it occurs in the southern Sierra Nevada and other southern mountains.

Acarospora obpallens, with a yellowish to dark brown thallus of flat, round or irregular areoles with small pits around the dark brown apothecia; spot tests C+ red, KC+ red, UV–; it is fairly common along the southern coast and in interior mountains. When very pruinose it can resemble *A. veronensis,* below.

Acarospora veronensis, with a thallus of brown, contiguous or dispersed areoles, sometimes very pruinose, with deeply immersed brown apothecia. Spot tests negative. A variable, fairly common species in southern CA, mostly coastal but occasionally found in the Sierra Nevada.

Amandinea punctata

Tiny button lichen

Thallus gray to brownish, thin and barely visible to thick, rimose to areolate and a bit scaly, becoming green when wet. Apothecia small, black, 0.2–0.5 mm wide, with thin rims that disappear when mature. Exciple uniformly dark brown. Spores brown, 2-celled, with uniformly thickened walls, 8 per ascus. Resembles a species of *Buellia,* but *Amandinea* spores are thin-walled, not constricted at the septum; conidia are short. Spot tests negative. Fairly common on wood or bark, rarely on rock or soil, statewide, but easily overlooked. The apothecia resemble those of *Catillaria nigroclavata* (not pictured), which has colorless spores; also compare with *Arthonia, Buellia, Bacidia, Lecidea,* and *Lecidella.* Photo is of a herbarium specimen from QC.

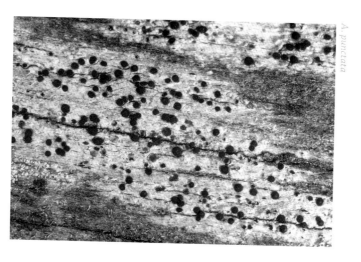

A. punctata

Arthonia
Comma lichens

A genus of crustose lichens with a rather heterogeneous collection of species. All have thin thalli, sometimes growing into the surface of the substrate, usually found on bark or wood, occasionally on rock or parasitic on other lichens. Apothecia vary in color but are often irregular or elongated, sometimes branching, immersed or flattened, usually without margins. Spores ellipsoid to fusiform, typically with one end wider than the other, 2–8-celled, thin-walled, cylindrical, usually 8 per ascus, often with 1 or 2 cells larger than the others, a characteristic of *Arthonia* (as well as *Opegrapha*). Most spot tests negative, with some exceptions described below. A notable species is *A. atra* (not pictured), which has a thin, pale grayish thallus, simple or branched apothecia, and black carbonaceous walls as in *Opegrapha* and unlike any other species of *Arthonia;* on smooth bark, coastal, San Mateo Co. to Mexico. Until recently it was called *Opegrapha atra*.

Arthonia cinnabarina
Bloody comma lichen

Thallus mostly immersed in bark but usually forming an orange or pinkish patch, sometimes with a narrow brown prothallus. Apothecia elongated, branched or even irregularly ring-shaped, and covered with deep red-orange pruina that make the lichen distinctive. Spores 5–8-celled. Spot tests: red pruina K+ purple; walls of hyphae I+ pale blue. On smooth bark in coastal areas, northern to central CA. Photo from Patricks Point State Park, north of Eureka.

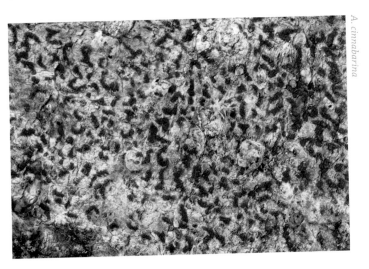

A. cinnabarina

Arthonia pruinata
Pastel comma lichen

Thallus pale greenish gray, thin and cracked. Apothecia thin, angular to rounded, reddish brown beneath a coating of white pruina, often appearing pale violet or pinkish. Spores persistently colorless, 3–6-celled. Spot tests: C– or C+ red, thallus and subhymenium I+ pale blue, ascomatal gels I+ blue turning red, KI+ blue. On bark in coastal zones, central to southern CA. Photo from the Channel Is.

A. pruinata

Aspicilia

Sunken disk lichens

A mostly crustose genus (a few species are fruticose) with many species, widespread and often common. Generally with flat or convex areoles and usually black apothecia that are flush with the thallus surface or often immersed, sometimes pruinose, and sometimes with a slightly raised lecanorine margin. Thallus can be white, gray, tan, or greenish, areolate to rimose, rarely continuous. Spores large, ellipsoid or subspherical, colorless, thin-walled, 1-celled, 4–8 per ascus; the conidia are relatively long, thin, and straight. Species of *Aspicilia* vary considerably in morphology and chemistry; they are found mostly on rock, some on soil. One species, *A. cyanescens,* can grow on bark or wood. Quite a few *Aspicilia* species found in CA are not pictured in this book; most are less common than those described here.

Aspicilia californica

Shrubby sunken disk lichen

Thallus small, fruticose, forming stringy whitish to gray filaments with simple tips that attach at several places along their length. Sometimes with black pycnidia. Apothecia rare. Spot tests: cortex and medulla K+ red, KC–, C–, P+ orange, I–. Occasional on moss, organic debris, or rock in central and southern low-elevation areas. Apparently endemic to CA. Another fruticose species, *A. filiformis* (not pictured), with an olive to brown thallus that reacts K–, is very rare in CA. Photo from a herbarium specimen.

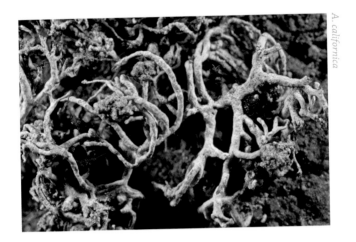

Aspicilia cinerea

Cinder lichen

Thallus variable, gray to off-white or greenish, continuous to areolate, with areoles separated by distinct cracks, sometimes with pycnidia. Apothecia black or very dark brown, usually flat or concave, and level with the thallus surface, sometimes with margins more or less thallus-colored, mostly epruinose. Spores smaller than most species in the genus. Spot tests: cortex and medulla K+ red, KC–, C–, P+ yellow, or sometimes medulla K+ persistent yellow, P+ orange. Rather rare in CA, on rock, mostly in mountains. Photo from WV.

Aspicilia confusa

Chaparral sunken disk lichen

Thallus with convex, rounded or irregular, sometimes dispersed areoles, usually pale gray but can be brownish, with a dark, fringed prothallus, sometimes with a thin, whitish pruina. Apothecia sunken, black, 0.1–1.5 mm wide, usually with 1, but up to 4, per areole. Spot tests negative. Quite variable, and rather similar to a number of other *Aspicilia* species not pictured in this book. On rock in chaparral or forest habitats, often in mountains, central and southern CA, including the Sierra Nevada, but not in southeastern deserts. Endemic to CA. Photo from near the coast, San Luis Obispo Co.

Aspicilia cuprea
Copper sunken disk lichen

Thallus areolate, often forming large patches up to 20 cm across, with flat or convex, tan to dark brown or grayish areoles; pycnidia common. Apothecia sunken, black, often thinly pruinose, usually with 1, but sometimes several per areole, and light rims that contrast with the dark disk. Spot tests: cortex and medulla K+ red, C–, P+ orange, I–. Fairly common on siliceous rock, in interior locations, central and southern CA. Photo from Mt. San Jacinto, Riverside Co.

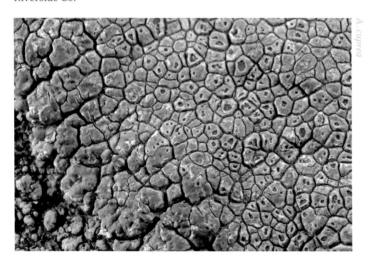

Aspicilia cyanescens
Bluish sunken disk lichen

Thallus areolate to rimose, rather rough, pale gray or with a bluish tinge. Apothecia black or slightly bluish, epruinose. Often on rock, but unique among CA species of *Aspicilia* in sometimes growing on bark or wood, especially that of incense cedar, sometimes on white fir or giant sequoia. Usually with a black to bluish or greenish prothallus when on bark; prothallus often absent when on rock. Spot tests negative. Central Sierra Nevada and southern mountains. Endemic to CA. Compare with *Megaspora verrucosa*. Photo of a specimen on granite from near Shaver Lake, Fresno Co.

A. cyanescens

Aspicilia pacifica
Pacific sunken disk lichen

Thallus areolate, with rounded or angular, flat to slightly convex areoles that are contiguous but separated by big cracks; white to grayish, brownish or ocher, often whitish at the edge and usually with a dark prothallus. Apothecia

A. pacifica

small, black, concave to flat, 0.1–0.8 mm wide, usually with white pruina and often with a thin grayish rim; immersed pycnidia common. Spot tests: cortex and medulla K+ yellow to red, C–, P+ orange, I–. Fairly common on coastal rocks along the seashore and higher in coastal mountains, Solano Co. to Mexico. Endemic to CA and Baja CA. Photo from San Luis Obispo Co.

Aspicilia phaea
Dusky sunken disk lichen

Thallus areolate (rarely rimose) with angular and irregular areoles, brown to grayish, sometimes with a dark prothallus at the edge. Apothecia common, black, usually concave, most often with 1 but up to 5 per areole, occasionally with white pruina, often with white to gray margins; with or without pycnidia. Spores comparatively large, colorless, simple, 8 per ascus. Spot tests negative. Fairly common on rock in coastal and inland areas, central and southern CA. Photo from Mt. San Jacinto, Riverside Co.

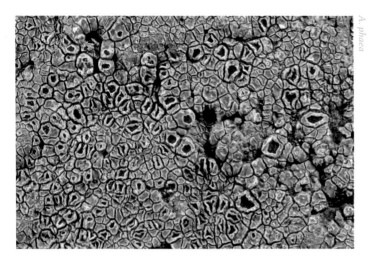

A. phaea

Bacidia heterochroa
Black dot lichen

Thallus thin, generally continuous but sometimes with cracks, white to pale grayish, sometimes with a narrow black prothallus. Apothecia irregular, flat black to purplish or mottled, with thin margins the same color as the disk or a bit paler; mostly epruinose. Spores straight or curved, colorless, 4–16-celled, 8 per ascus. Spot tests negative. On bark of broadleaved trees in coastal areas, Sonoma Co. to Mexico. Compare with *Amandinea, Buellia, Lecidea,* and *Lecidella.* A few other species of *Bacidia,* not pictured, occur in CA, as well as *Bacidina californica* (not pictured), a crustose lichen with a pale gray to greenish thallus, often sorediate, with small, biatorine apothecia varying in color from pink to yellow to brownish or black; spores colorless, 4–10-celled (rarely 1-celled), 8 per ascus; found occasionally on bark in coastal habitats from Sonoma Co. to the Channel Is. Photo from the Coast Range near Santa Barbara.

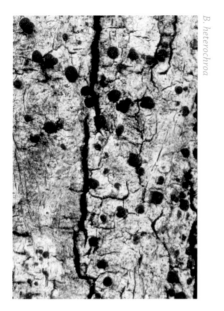

B. heterochroa

Bellemerea alpina

Brown sunken disk lichen

Crustose thallus, off-white to pale tan, rather thick, areolate, with continuous or dispersed areoles, often with a conspicuous black prothallus; epruinose. Apothecia pale or darker, brownish to gray or a bit purplish, embedded in areoles, often pruinose. Resembles a species of *Aspicilia*, especially when the apothecia are more black than brown. Asci like those of *Porpidia*. Spores broadly ellipsoid, 1-celled, 8 per ascus. Spot tests: thallus K+ yellow turning red, KC–, C–, P+ deep yellow. A mountain lichen, occasional on rock in the Sierra Nevada, rare in southern mountains. Photo from interior BC.

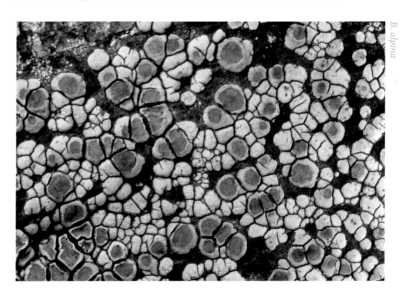

B. alpina

Buellia

Button lichens

Crustose lichens with white to gray, yellowish or brown thalli that can be thin or thick, continuous to rimose or areoleate. Apothecia black, lecideine, with black margins (but sometimes margins lacking) on the surface or immersed between areoles, sometimes pruinose. Spores brown, 2-celled (rarely 4-celled) or muriform, often constricted at the septa 4–8 (8–16) per ascus. On rock, bark, wood, soil, or parasitic on other lichens in many habitats. Compare with *Amandinea, Bacidia, Catillaria, Diplotomma, Lecidea,* and *Lecidella.*

Buellia badia

Parasitic button lichen

Thallus dark brown, more or less squamulose. Apothecia black, lecideine. Rarely pruinose, without soredia. Spores oblong to ellipsoid with obtuse ends, straight, brown, 2-celled, typically 10.5–13 × 6–7.5 µm, 8 per ascus. Spot tests negative. Initially grows parasitically on a variety of other lichens, as on *Aspicilia phaea* in the photo, eventually becoming independent on rock or, rarely, on hard wood. The photo also shows some sort of pathological condition on the *Buellia*, visible in the yellow and reddish decayed portions. Closely resembles *Dimelaena californica* in both outward appearance and internal characteristics; *D. californica,* which reacts K+ red in the medulla, usually gets its start by growing on *Dimelaena radiata.* Fairly common Mexico to Sonoma Co., Coast Range and southern Sierra Nevada. Photo from Marin Co.

B. badia

Buellia concinna

Cinnabar button lichen

Thallus granular to minutely bullate, pale yellow to greenish or a bit brown, without a prothallus. Apothecia black, lecideine, with distinct margins. Spores ellipsoid with tapered ends, often curved, brown, 2-celled, typically

15.6–21.3 × 7.9–9.9 µm, 8 per ascus. Spot tests: usually K+ yellow, KC–, C+ orange or pinkish, P+ orange, CK+ orange, UV+ yellow to orange. Moderately common on siliceous rock, especially cliff faces, statewide, but especially in the Sierra Nevada and southern mountains. Photo from Mt. San Jacinto.

B. concinna

Buellia disciformis
Boreal button lichen

Thallus thin, pale gray to bluish, smooth or cracked to areolate, epruinose. Apothecia black, flat, up to 1 mm wide, with distinct lecideine margins. Oil drops can be seen in hymenium. Spores ellipsoid, typically 18–26 × 6–11 µm, 8 per ascus. Spot tests negative except K+ yellow. On bark, sometimes on wood, in forests, mostly coastal, statewide. A similar species, *B. erubescens* (not pictured), has smaller spores, reacts K+ yellow to orange-red, and is more often found on wood. Photo from northern ME.

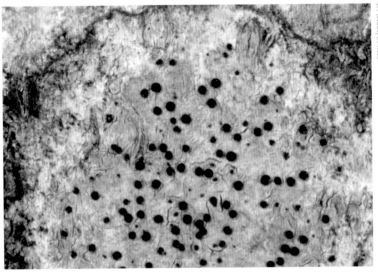

B. disciformis

Buellia dispersa

Scattered button lichen

Areolate thallus, with dispersed areoles that can become subsquamulose, or forming rosettes with lobate margins; usually ivory but can be brown to gray, generally not shiny and with some pruina. Apothecia black, lecideine, usually epruinose. Spores narrow, oblong to ellipsoid with obtuse ends, not curved, brown, 2-celled, typically 12–16 × 6.5–8.5 µm, 8 per ascus. Spot tests: K+ yellow, KC–, C–, P+ yellow. Fairly common on rock in both coast and inland areas, central and southern CA. Resembles *B. nashii* (not pictured), which has a white, K+ red thallus, black, usually epruinose apothecia, and blue-green pigments in the epihymenium and exciple; also on rock, southern Coast Range and Sierra Nevada. Also similar is *Diplotomma venustum* (not pictured), with apothecia that usually have a thalline collar and 4-celled to submuriform spores, found in southern coastal and inland areas. Photo from the Mojave Desert.

B. dispersa

Buellia halonia

Seaside button lichen

Thallus pale yellowish green, rimose to areolate, continuous, and usually forming distinctive round patches a few cm across. Apothecia black, immersed when young, becoming superficial, 0.4–1.2 mm wide, slightly convex and with bluish pruina. Spores polarilocular, dark green to brown, 11–17 × 6–10 µm, 8 per ascus. Spot tests: cortex K+ yellow, KC+ red-orange, C+ orange, CK+ orange, P–; medulla IKI–. Fairly common and usually easy to recognize, although it could be confused with the rarer *Lecidella asema*, which has colorless, 1-celled spores. On coastal rocks, Sonoma Co. to Mexico. Photo from the Channel Is.

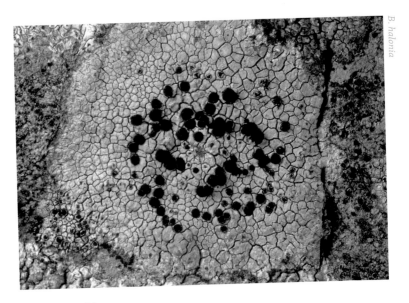

Buellia maritima

Coast rock button lichen

Thallus whitish, thin, typically growing in distinct round patches and often with adjacent thalli crowding together, rimose to rimose-areolate, heavily pruinose. Apothecia tiny, black. Thallus contains calcium oxalate crystals. Spores narrow, oblong to ellipsoid with obtuse ends, not curved, brown, 2-celled, typically 10.4–12.8 × 5.1–6.5 µm, 8 per ascus. Spot tests: thallus usually K+ yellow to red, C–, KC–, CK–, P+ yellow. Occasional on coastal rock, especially rock containing calcium carbonate, central to southern CA. Photo from Santa Rosa Is.

Buellia oidalea
Coast wood button lichen

Thallus creamy to pale gray or yellowish, thin and rimose to areolate or thicker and rugose to verruculose, or even subsquamulose, often with a black prothallus. Apothecia black, 0.2–2 mm wide, flat at first, then becoming convex, with lecideine margins that can disappear with age, epruinose. Spores oblong to ellipsoid with obtuse ends, colorless to olive or brown, with 13–40 cells; typically 34–45.5 × 13.5–18 μm, usually 8 per ascus. Spot tests: thallus K–, C+ orange, P–; medulla no reactions. Characterized by large, muriform spores that soon become brown and have unevenly thickened walls. Fairly common on bark and wood in coastal areas, central to southern CA. Farther north, from Monterey Co. to BC, the rarer *Buellia muriformis* (not pictured), which occupies much the same habitat, has a whitish gray K+ yellow, C– thallus, and its spores become brown only in age and have no spore wall thickenings. Also resembles *Diplotomma penichrum*, which has a whiter, K+ yellow, C– thallus, much smaller spores, and a more montane habitat. Photo from a herbarium specimen collected in CA.

B. oidalea

Also of Note

A few other, fairly common *Buellia* species, not pictured, are:

Buellia christophii, with a brown thallus and black apothecia, epruinose; spores broadly ellipsoid to nearly spherical, colorless, typically 10.5–13 × 7–8.5 μm; spot tests negative; fairly common on coastal rock in central and southern CA.

Buellia ryanii, with an olive gray to brownish thallus; apothecia black, epruinose; spores broadly oblong to ellipsoid, soon brown, typically 10.5–12.2 × 5.6–6.8 μm; spot tests negative; on rock in central to southern coastal areas.

Buellia sequax, with a pale brown to gray thallus; apothecia black, epruinose; spores oblong becoming ellipsoid, soon brown, typically 10.7–12.8 × 3.8–5.4 μm; spot tests negative; on oak bark in southern coastal areas; formerly called *B. lepadistroidea*.

Buellia spuria, with a white to gray thallus; apothecia black, epruinose; spores broadly oblong to ellipsoid, soon brown, typically 11.9–14.8 ×

5.9–7.4 µm; spot tests: K+ yellow to red, C–, KC–, P– or P+ yellow; fairly common on rock in the mountains and desert.

Buellia tesserata, with a white to pale gray thallus, not shiny; apothecia black, often pruinose; spores oblong becoming ellipsoid, soon brown, typically 9–10.6 × 4.6–5.7 µm; spot tests: K+ yellow, rarely red, C– or rarely fleeting pink, KC–, P– or P+ faint yellow; common on coastal rocks, Marin Co. to Mexico.

A number of other, less common, species of *Buellia,* not pictured, occur in CA.

Calicium
Stubble lichens, Pin lichens
Crustose thalli, powdery or granular to thin and barely visible. Apothecia are tiny unbranched black stalks, usually less than 2 mm tall, with cuplike caps ("capitula") containing dark brown spores in a black mass called a mazaedium; sometimes the stalks or capitula are coated with white or yellow pruina. On wood and bark in shady forests. Spores dark brown, 2-celled, with a wall between the cells. Stubble lichens are hard to spot because of their tiny size; often the best way to find them is to look at the edge of a dead tree trunk in silhouette. They have been used as indicators of old-growth forests and ancient trees. Compare with *Chaenotheca*. A recently discovered species, *Calicium sequoiae* (not pictured), has been found on old-growth redwoods in northwestern CA. It has white-pruinose apothecia and unusual spores with spiral ridges. The thallus reacts K+ yellow, P+ orange; stalks react I+ blue.

Calicium glaucellum
White ring stubble lichen
Thallus usually immersed in the substrate and not visible, otherwise thin, dark gray-green and granular; caps on the short, thick stalks have a white ring of pruina around the lower part of the capitulum. Spot tests: thallus K+

C. glaucellum

dull yellow, KC–, C–, P–. On wood, coast to mountains, probably widely distributed but easily overlooked. Photo from Wells Gray Park, BC.

Calicium viride

Green stubble lichen

Thallus granular green to yellow-green with slender stalks up to 2.5 mm high bearing ball-like, black, epruinose capitula that are brown underneath. Spot tests negative except UV+ orange. Fairly common on wood and bark, mostly in mountain forests, occasionally on the coast; statewide but more frequently in northern CA. Photo from Yosemite National Park.

Caloplaca

Firedot lichens, Jewel lichens

A large group of crustose lichens with thalli in a number of colors, but most are orange or yellow; they can be continuous to areolate, subfruticose, squamulose, or not visible at all. Apothecia are usually orange, but also red, brown, or other colors in some species; most often lecanorine, but biatorine or lecideine in a few species. Spores colorless, 2-celled, usually with a wide septum (polarilocular), 8 per ascus for the species described below. Orange pigments in the thallus and apothecia (anthraquinones) usually react K+ deep purple. On all substrates, widely distributed. Species with marginal lobes may be confused with appressed *Xanthoria,* such as *X. elegans,* but *Xanthoria* species have a lower cortex and can be separated from the substrate without damage; to remove *Caloplaca* intact one has to take some substrate with it. Other species resemble *Candelaria* or *Candelariella,* but those are more yellow than orange. Because of their color, species of *Caloplaca* are among the most conspicuous lichens in the landscape, sometimes covering large patches of rock. Many species, not pictured, occur in CA; most are less common than the ones described below.

Caloplaca arenaria

Granite firedot lichen

Thallus growing into the rock and barely visible. Apothecia red-orange to brownish, 0.2–0.5 mm wide, flat and somewhat angular, with thin rims that are usually a bit paler than the rest of the disk; epruinose. Spores narrowly ellipsoid to almost cylindrical, typically 10–15 × 3.5–5.5 µm with narrow septa usually less than 2 µm thick. Most often in arid regions, southern CA or east side of the Sierra Nevada, but sometimes in coastal habitats. Resembles *C. luteominia* var. *luteominia*, but that species has browner apothecia and a more visible thallus. Photo from the Rocky Mountains, CO.

C. arenaria

Caloplaca bolacina

Waxy firedot lichen

Thallus bright yellow-orange, a bit waxy in appearance, often scattered into areoles or forming slightly lobed squamules, shiny or pruinose. Apothecia large, lecanorine, 0.7–2 mm wide, with darker orange disks and margins similar to the thallus in color. Spores ellipsoid, 12.5–17.5 × 5.5–8.5 µm with septa 3–5.2 µm thick. Compare with *C. stantonii*. Fairly common on rock, often in the coastal spray zone, along the entire length of the CA coast and in coastal mountains. The bright color and large apothecia make it one of the more conspicuous *Caloplaca* species. Photo from the Channel Is.

Caloplaca brattiae

Coast lobed firedot lichen

Thallus orange, sometimes a bit yellowish, with strongly convex lobes 0.5–2.5 mm long and 0.2–0.4 mm wide at the margin. Apothecia numerous, deeper orange, tending to cluster in the middle. Spores ellipsoid, typically 10–13 × 4–5 µm with septa 2–3 µm thick. Fairly common on coastal rock, Sonoma Co. to Ventura Co. Compare with *C. impolita*. Photo from Estero Bluffs State Park, San Luis Obispo Co.

237

C. cerina

Caloplaca cerina
Gray-rimmed firedot lichen
Thallus pale to dark gray or bluish gray, thin or more pronounced, usually continuous, smooth to areolate. Apothecia bright orange or yellow, lecanorine, with gray rims. Spores ellipsoid, 10–17 × 6–8.5 μm with septa 3–8 μm thick. Fairly common on deciduous tree bark in open areas, mountains and coast, statewide. A similar species, *C. stanfordensis* (not pictured), has a completely gray thallus with white to gray apothecial margins that react K–, C–; common on oaks. Photo from the Coast Range, San Luis Obispo Co.

Caloplaca citrina
Mealy firedot lichen
Thallus areolate to granular, strongly sorediate, dark yellow to yellowish orange. Fine yellow soredia are produced in laminal or marginal soralia; apothecia rare. The yellowish color and granular thallus make it somewhat resemble a leprose crust or a species of *Candelariella*, but the K+ red-purple spot test will identify it as a *Caloplaca*. Occasional on bark and on wood such as fence rails, both coast and mountains, statewide, but more frequent in southern CA. Photo from coastal MA.

Caloplaca coralloides
Coral firedot lichen
Thallus subfruticose, yellow-orange, with lumpy, short, terete, forking branches that can lie flat on the rock or form little raised cushions. Apothecia deeper orange, lecanorine, either abundant or absent. The fruticose growth

form resembles *Teloschistes,* but *C. coralloides* grows on rock close to the ocean, usually in the spray zone. Also resembles *Xanthoria candelaria,* but that species is minutely foliose and sorediate. A lichen of a different genus, *Edrudia constipans* (not pictured), found only on the Farallon Is., also resembles *C. coralloides.* Strictly coastal, statewide. Photo from Monterey Co.

Caloplaca crenulatella
Western sidewalk firedot
Endolithic, with no visible thallus; visible as small, adnate, orange, lecanorine apothecia 0.3–0.8 mm wide. Spores ellipsoid, typically 15.5–18 × 5.5–7 μm with septa 1–3 μm thick. Usually on calcareous rock in coastal and inland areas, most often in southern CA. Resembles *C. arenaria* in having a very

narrow spore septum and a largely invisible thallus, but that species grows on siliceous rock. Photo from the Granite Mountains Reserve, Mojave Desert.

Caloplaca decipiens
Deception firedot lichen

Thallus yellow-orange and pruinose, with radiating, convex lobes that touch along the sides and have well-defined margins; with soralia containing fine yellow soredia, especially on the lobe tips. Apothecia rare. Spores ellipsoid, 12.5–15.5 × 5.5–7 μm with septa 1.5–3 μm thick. On rock, statewide, in mountain areas. Photo from Zion National Park, UT.

Caloplaca epithallina
Parasitic firedot lichen

This species lives on other lichens with its thallus inside the host and not visible; all that can be seen of the *Caloplaca* are clusters of rusty red apothecia

on the surface of the host lichen (in the photo, a lobate species of *Lecanora*). Apothecial margins reddish to black, reacting C+ red. Spores broadly ellipsoid, 18–13 × 5–8 μm with septa 2–3.5 μm wide. Occasional on lichens growing on rock, Lake Tahoe to southern CA. Photo from the Rocky Mountains, CO.

C. epithallina

Caloplaca ignea
Flame firedot lichen

Thallus brilliant red-orange, often areolate toward the center, with elongated, narrow lobes around the margin; perimeter often slightly paler in color. Apothecia 0.3–1 mm wide, the same color as the thallus or slightly lighter, with margins sometimes a bit lighter than the center. Spores ellipsoid, typically 11–12.5 × 5.5–7 μm with septa 2–3 μm thick. Compare with *C. saxicola*. On rock in coastal and inland habitats, scattered statewide. In the central

C. ignea

Sierra Nevada foothills or on rock outcrops in the Coast Range, this species can be a real eye-catcher! Photo from Merced Co.; the yellow lichen in the photo is probably *Acarospora socialis*.

Caloplaca impolita
Fan-lobed firedot lichen

Thallus yellowish orange with marginal lobes 1.5–2 mm long, 0.3–1 mm wide. Apothecia orange, 0.3–1 mm wide. Spores ellipsoid, 12.5–14 × 5.5–7 µm with septa 4–5.5 µm thick. Resembles *C. brattiae*, but *C. impolita* is yellower, with lobes that are broader and flatter at the tips, and it is frequently pruinose. On noncalcareous coastal rock, Sonoma Co. to Mexico. Photo from Mt. Tamalpais, Marin Co.

C. impolita

Caloplaca inconspecta
Seaside firedot lichen

Thallus may be scattered to areolate, yellow-orange to yellow, or almost invisible. Apothecia bright yellow-orange, often in patches. Spores broad, small, typically 10–14 × 3.5–7.5 µm with septa typically 3–5 µm wide. Uncommon, on coastal rock in the upper intertidal zone or on driftwood, central CA coast. Photo from Marina Park, Berkeley.

Caloplaca ludificans
Firedot lichen

Thallus with little apparent structure, appearing simply as a gray to yellowish color on the rock or sometimes continuous to areolate and thinning at the edge. Apothecia adnate, deep orange, with lecanorine margins or no margins at all. Spores ellipsoid, typically 12.5–17 × 5.5–7 µm with septa 1–3 µm wide. On rock, almost always coastal, statewide. Closely resembles *C. luteominia* var. *luteominia*, but the thallus in *C. ludificans* is smooth, with a translucent, waxy

appearance, and the apothecial margins are thinner and flatter or absent. Photo from Marina Park, Berkeley.

Caloplaca luteominia
Red firedot lichen

This species has two varieties, var. *bolanderi*, with apothecia a brilliant rose-red, unlike any other lichen, and var. *luteominia*, with orange-brown apothecia. The thallus of both is often entirely within the rock or barely visible on the surface, areolate or continuous, pale gray or brown to reddish. Apothecia are 0.4–1.2 mm wide, with lecanorine margins. Spores ellipsoid, typically 14–20 × 4–7.5 μm with septa 2–3 μm thick. Both varieties are occasional along the coast and in coastal mountains, central to southern CA. Var. *bolanderi* is endemic to CA. Compare var. *luteominia* to *C. arenaria* and *C. ludificans*.

Caloplaca luteominia var. *bolanderi*. Photo from Alameda Co.

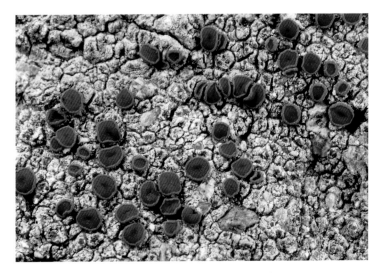

Caloplaca luteominia var. *luteominia*. Photo from the Channel Is.

Caloplaca marina subsp. *americana*
Marine firedot lichen

Thallus areolate, yellowish, without lobes, sometimes poorly developed. Apothecia crowded, lecanorine, more or less the same color, 0.3–1 mm wide. Spores ellipsoid, typically 12.5–14 × 5.5–7 µm with septa 3–4 µm thick. Resembles *C. rosei*, but the thallus of *C. marina* is more yellowish and discontinuous, with more convex areoles and more concave, crowded apothecia with prominent margins. *Caloplaca marina* subsp. *marina* occurs in Europe. Photo from rocks at Marina Park, Berkeley; the small fruticose thalli in the picture are *Xanthoria candelaria*.

Caloplaca rosei
Rose's firedot lichen

Thallus continuous to areolate, apricot-orange, without elongated lobes; the margin becomes thin around the edge. Apothecia lecanorine, more or less the same color, without pruina. Spores ellipsoid, typically 11–14 × 5.5–7 μm with septa 3–5.5 μm thick. Compare with *C. marina*, which has a more minutely areolate, almost granular thallus, and no prothallus. The two species often grow together, which can confuse attempts at identification. On coastal rock, entire length of coast. Photo from Salt Point State Park, Sonoma Co.

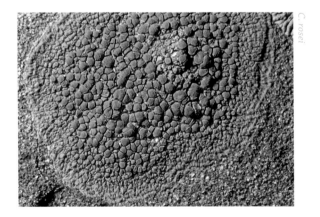

Caloplaca saxicola
Rock firedot lichen

Thallus small, with elongated, short, somewhat inflated-looking lobes that have an abrupt edge, bright orange. Apothecia adnate or immersed, the same color as the thallus, often developing near the lobe tips. Spores ellipsoid, typically 11–14 × 5.5–7 μm with septa 2–4 μm thick. Resembles *C. ignea,* but that species has larger thalli with apothecia typically clustered near the center. One of the more common rock species, on the coast and in the mountains, statewide. Photo from Joshua Tree National Park.

Caloplaca stantonii
Bicolor firedot lichen

Thallus yellowish, areolate to squamulose, with a slightly lobed margin. Apothecia brownish orange, with small or no margins, epruinose. Spores ellipsoid, typically 15.5–19.5 × 5.5–7 μm with septa 1.5–3 μm thick. Resembles *C. bolacina,* but the thallus is yellower and the apothecia are smaller and darker, with less prominent margins. Mostly, but not exclusively, coastal, Sonoma Co. to Orange Co. Photo from the Channel Is.

Caloplaca stipitata
Stipitate firedot lichen

Thallus areolate, yellowish, smooth or verruculose, without elongated lobes, often with a black prothallus. Apothecia flat, orange, stipitate (stalked), with slightly raised margins the same color as the thallus. Spores ellipsoid, typically 12.5–14 × 5.5–7 µm with septa 4–5.5 µm thick. Resembles *C. californica* (not pictured), a rarer, more northern, maritime species on wood that has crowded, nonstipitate apothecia. On bark in coastal areas, southern CA. Photo from Baja CA.

Caloplaca subsoluta
Loose firedot lichen

Thallus areolate to somewhat squamulose, without elongated lobes but often with a slightly notched but not abrupt margin; bright orange, often with a black prothallus. Apothecia red-orange, adnate, lecanorine. Spores ellipsoid, typically 9.5–12.5 × 5.5–7 µm with septa 3–4 µm thick. Resembles *C. squamosa* (not pictured), but that species has a more squamulose thallus, thicker and more convex lobes, and apothecia with prominent thalline margins; central and southern Coast Range and interior. Fairly common on rock in coastal and mountain areas, statewide except for the north coast. Photo from Tonto National Forest, AZ.

Also of Note

Not pictured, but worth noting because they are fairly common, are:

Caloplaca albovariegata, with a bluish gray thallus; apothecia black with gray margins; spores ellipsoid, typically 14–18 × 7–10 μm with septa 1.5–4 μm; on rock on the coast and in the interior, Marin Co. to Mexico.

Caloplaca atroflava, with a dark gray or brownish thallus; apothecia dark orange with mostly orange margins or with an outer gray rim; spores ellipsoid, typically 11–17 × 5.5–8.5 μm with septa 4.5–5 μm; on rock, statewide, Coast Range.

Caloplaca demissa, with an olivaceous gray to brown thallus and marginal, short to elongated lobes, without a prothallus; apothecia absent but with granular soredia in laminal soralia; fairly common on vertical sunny rocks statewide, especially in mountains. Can somewhat resemble *Koerberia sonomensis,* but that is a cyanobacterial lichen, and foliose.

Caloplaca ferruginea, with a gray thallus; apothecia dark, rusty red-orange with orange margins; spores ellipsoid, typically 12.5–17 × 7–10 μm with septa 4–7 μm; on wood or bark, statewide, mostly coastal.

Caloplaca nashii, with a largely invisible thallus immersed in calcareous rock; apothecia orange; spores ellipsoid, typically 12.5–14 × 4–5.5 μm with septa 1–1.5 μm; in the central and southern mountains, coast and inland.

Caloplaca pyracea, with a thin, areolate thallus, sometimes growing within the bark and barely visible; gray to whitish, often with some pale orange portions; apothecia orange, 0.3–0.5 mm wide with lighter orange margins; epruinose; spores ellipsoid, typically 10–14 × 5.5–7 μm with septa 1.5–4 μm thick; spot tests: apothecial margins K+ red; statewide, more often found in coastal southern CA but in mountains and farther north as well. Resembles *C. crenulatella,* but that species grows on calcareous rock and has narrower spore septa. *C. pyracea* is a newer, tentative name for what used to be called *C. holocarpa* and may include more than one species.

Caloplaca stellata, with an orange thallus that typically has star-shaped, radiating lobes bearing fine soredia in soralia at the lobe tips; apothecia rare; spores ellipsoid, typically 11–12.5 × 5.5–7 μm with septa 3–4 μm; fairly common on rock, statewide.

Candelariella

Goldspeck lichens, Yolk lichens

Bright yellow, ocher, or greenish yellow, crustose or squamulose lichens, often areolate or granular, sometimes with little visible thallus. Apothecia are yellow, lecanorine, and not sunken, as in yellow species of *Acarospora* or *Pleopsidium.* Few species of *Caloplaca* are bright yellow. Spores ellipsoid, colorless, 1-celled (rarely 2-celled), 8–32 per ascus. Spot tests: K– or K+ pale rose, KC–, C–, P–, UV+ dull, dark orange. On all substrates, widespread and common. Most yellow crustose lichens that one finds on rock are species of *Candelariella;* orange crusts are usually *Caloplaca.*

Candelariella aurella

Hidden goldspeck lichen

Thallus of small, scattered, dark yellow areoles or granules, often barely visible. Apothecia numerous, small, 0.25–1 mm wide, margins often thin or absent. Spores 8 per ascus. Compare with *C. vitellina,* with 16–32 spores. Common on calcareous rock, sometimes on wood or bark, widespread in coastal and mountain areas. Often scattered in patches on concrete in urban locations, looking rather like spots of yellow paint but (unlike paint) turning greenish after a rain. Photo from western CO.

Candelariella citrina

Tundra goldspeck lichen

Thallus bright greenish yellow to somewhat orange-yellow, granular to areolate or developing lobate squamules, sorediate at edges of squamules or areoles with soredia sometimes taking over the entire thallus. Apothecia same color as the thallus, 0.3–0.8 mm wide. Spores short, with one or both ends pointed, 8 per ascus. On soil, sometimes on rock, especially in small cracks, or on moss or dead vegetation, in coastal and mountain areas, central to southern CA. Photo from southern ID.

Candelariella rosulans
Sagebrush goldspeck lichen

Thallus dark yellow with lobed squamules, sometimes raised above the substrate, or scattered convex areoles. Apothecia 0.4–2.2 mm wide with thick, raised margins, disks slightly darker than the thallus. Spores 8 per ascus. In open, arid habitats, usually on rock but occasionally on wood or bark, mostly inland, especially in mountains. One of the most common species of *Candelariella*. Photo from San Luis Obispo Co.

Candelariella vitellina
Common goldspeck lichen

Thallus of small yellow areoles with crenulate margins, looking almost granular. Apothecia common but sometimes absent. Spores 16–32 per ascus. Spot tests: K+ reddish. Widespread and common on granitic rock, also on wood, rarely on bark, on the coast and in the interior, statewide. Resembles *C. aurella* but grows on noncalcareous rock. If on bark, consider *C. lutella,* mentioned just below. Photo from Napa Co.

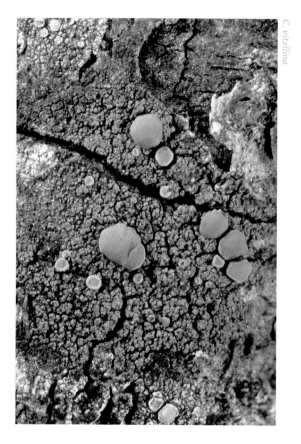

C. vitellina

Candelariella xanthostigma
Scattered goldspeck lichen

Thallus of yellow, almost spherical, corticate granules, 0.15–0.7 mm wide, often scattered but sometimes clustered together. Apothecia usually absent or sparse, rarely abundant. Spores 16–32 per ascus. Spot tests: K+ reddish. Widespread but not especially common, on bark of broadleaved trees, sometimes on conifer bark or wood, coast and interior. A similar species is *C. lutella* (not pictured), which, like *C. xanthostigma,* is minutely areolate, but with abundant, tiny apothecia containing many spores per ascus; occasional on bark statewide. Photo from a herbarium specimen collected in ON.

Catillaria lenticularis
Lens dot lichen

Thallus crustose, white to tan, usually indeterminate and somewhat immersed in the substrate. Apothecia brown to blackish, more or less flat, 0.2–0.7 mm wide, lecideine, epruinose, with thin or absent margins. Spores colorless, 2-celled, 8 per ascus. Spot tests negative. Occasional on limestone or other calcium-rich substrates, usually coastal, southern CA to at least Sonoma Co. Resembles a species of *Buellia* or *Amandinea;* when the thallus is mostly immersed, it could be mistaken for an endolithic species of *Lecidea,* such as *L. laboriosa,* but more often the thallus is somewhat visible. Other species of *Catillaria,* not pictured, include *C. chalybeia,* with a dark grayish to olive thallus, almost black, convex apothecia, found on acidic rock (rarely on bark) especially near lakeshores and the coast, and *C. nigroclavata,* with an indistinct, grayish, granular thallus and dark apothecia, found on bark or wood in the Coast Range; it resembles *Amandinea punctata,* but that species has brown spores. Photo from Salt Point State Park, Sonoma Co.

Chaenotheca furfuracea
Sulphur stubble lichen

One of several *Chaenotheca* species in CA, this one is distinctive for its bright yellow-green color over the entire lichen. Its minute stalks grow only 2–4 mm high, each with a rounded, sulfur yellow head, the capitulum, within which is a mazaedium, a brown mass of spores, spherical, brown to nearly colorless, 1-celled. On soil and wood, especially on the exposed roots of overturned old stumps, rarely on rock. The rarer *C. chrysocephala* (not pictured) has a yellow, pruinose thallus with a yellow ring around the top of an otherwise brown capitulum; the capitula of other species are browner to black. Compare with *Calicium*. Photo from Marin Co.

C. furfuracea

Chrysothrix
Gold dust lichens

Leprose thalli consisting of powdery soredia, rarely with apothecia, and without other visible structures. Color is a brilliant yellow or yellow-green, the brightest of any lichen, sometimes with tiny flecks of orange. Individual soredia are very small, less than 0.1 mm wide. *Chrysothrix* species usually form irregular patches on shady bark, on wood such as fences, or on rock, sometimes covering large areas. Differentiating among species depends on details of tissue structure and chemistry.

Chrysothrix granulosa
Coastal gold dust lichen

Thallus forming vague cushions of soredia over a fluffy medullary layer, brilliant yellow. Apothecia absent. Spot tests: K+ orange, C-, KC-, P+ orange; UV+ dull dark orange. Usually on bark, occasional on wood or rock, mostly coastal, statewide. Photo from Contra Costa Co.

Chrysothrix xanthina

Sulphur dust lichen

Thallus brilliant yellow to yellow-green or slightly orange, powdery, thin, lacking medullary tissue. Apothecia extremely rare. Spot tests: K–, C–, KC–, P–or P+ orange; UV– or UV+ dull orange. Common on bark, wood, and rock in dry, shady habitats, statewide, most frequently coastal, but sometimes inland and in mountains. Photo from northern WA.

Circinaria arida

Pebble ball lichen

Thallus brown to olive, areolate to verrucose, with rounded, convex areoles forming a cluster of tiny balls, each one with a sunken black, but pruinose, apothecium with a white rim; sometimes with immersed pycnidia. Spores simple, rounded, colorless, 14–36 × 13–28 µm. Spot tests: cortex and medulla I–, K–, P–, C–. Compare with *Circinaria contorta;* it could also be mistaken for a species of *Aspicilia.* On rock, especially small pebbles, in arid habitats but occasionally coastal areas, central to southern CA. Photo from the rim of the Grand Canyon, AZ.

C. arida

Circinaria contorta

Chiseled sunken disk lichen

Thallus with thick, rounded, olive gray or greenish areoles that have a chiseled look, continuous or dispersed, frequently with white, depressed

C. contorta

pseudocyphellae. Apothecia deeply immersed, mostly one per areole, pruinose. Spores broadly ellipsoid to spherical, 20–30 × 15–25 μm, 4 per ascus. Spot tests negative. Occasional on rock in southern deserts.

Clavascidium lacinulatum
Brown stipplescale

Thallus consisting of squamules, usually separated but occasionally overlapping, most often dark brown, but sometimes reddish or almost black, turning greenish when wet. Lobes have perithecia, appearing as black dots; epruinose. Lower surface tan, with pale rhizines and clumps of hyphae that attach to the substrate. Fairly common on soil in deserts and mountains, coast and inland, central to southern CA. Resembles species of *Placidium,* especially *P. lachneum,* which has a black lower surface and a lower cortex with vertical columns of cells, but species of *Placidium* lack rhizines. Photo from Organ Pipe Cactus National Monument, AZ.

C. lacinulatum

Cliostomum griffithii
Multicolored dot lichen

Thallus usually thin, pale gray or with a green, blue, or yellow tint, continuous or with some cracks, and sometimes mostly inside the substrate so not visible, particularly when on wood. Apothecia small, flat to slightly convex, adnate, varying in color, often on the same thallus, from dark gray to pinkish or beige, with biatorine margins that may be thin, thick, or absent. Often with abundant, black, irregularly shaped pycnidia. Spores colorless, 2-celled, occasionally 4-celled, often with oil droplets, 8 per ascus. Spot tests: K+ yellowish (check under a microscope), KC–, C–, P–. Small and easily overlooked, but distinctive because of the multicolored apothecia. Occasional on bark and wood in coastal habitats statewide. Photo from NB.

Coenogonium luteum

Orange dimple lichen

Thallus thin, crustose, greenish gray, almost like a coating of varnish, continuous or cracked. Apothecia scattered, biatorine, sessile, pale yellow to pink-orange, constricted at the base, epruinose. All spot tests negative. Spores elongate ellipsoid, colorless, 2-celled, 8 per ascus. Grows on all substrates, often over moss, in shady, moist habitats, mostly in coastal forests, statewide. An uncommon but distinctive lichen. Photo from a herbarium specimen, collected in QC.

Collemopsidium halodytes
Barnacle lichen

Thallus thin and membranous, brown, often scattered, with perithecia that typically have a black ring around the ostiole. Photobiont is *Hyella*, a cyanobacterium. Spores ovate, 1-celled, uneven in size, 8 per ascus. Sometimes on coastal limestone, but usually seen on seashells such as barnacles and limpets in the littoral zone along the entire coast. The photo, from a herbarium specimen collected in coastal BC, shows the lichen on a barnacle.

Cyphelium
Soot lichens

Crustose lichens with thin, gray or yellow areolate thalli and apothecia that are cuplike mazaedia with a mass of sooty black spores. Mazaedia have a black exciple and may be immersed or prominent. Spores dark brown to black, 2-celled or rarely submuriform. On hard, dead wood, or hard conifer bark. A few other species, not described below, occur in CA, of which only *C. tigillare* is relatively common.

Cyphelium brachysporum
Chaparral yellow soot lichen

Areolate to verrucose thallus, yellow to yellow-green, sometimes thin to almost completely immersed in the substrate. Mazaedia black, containing 2-celled to submuriform, almost spherical spores are imbedded in the areoles, contrasting with the yellow thallus surrounding them. Rare, found on unburned wood in old-growth chaparral; it looks almost identical to the more common and widespread *C. tigillare*, which has larger, 2-celled ellipsoid spores. Both species are usually found on wood, occasionally on conifer bark, in coastal habitats, central to southern CA. Photo from San Luis Obispo Co.

Cyphelium inquinans

Cupped soot lichen

Thallus grayish, verrucose to areolate, thick and well developed or sometimes thin. Apothecia black, usually sessile, not immersed. Exciple edge often lightly pruinose, creating a light ring around the inner disk. Spores constricted in the middle. Spot tests: K+ yellow to brownish, KC– or slightly orange, C–, P– or P+ pale yellow to reddish. Compare with *Amandinea, Bacidea, Buellia, Lecidea,* and *Lecidella*. On conifers or dead wood in shady, moist habitats, scattered statewide but easily overlooked. Photo from interior BC.

Cyphelium pinicola
Pine soot lichen

Thallus verrucose, bright greenish yellow, sometimes almost immersed in the substrate. Apothecia convex, completely black, shiny, epruinose, sessile rather than immersed as in *C. tigillare* or *C. brachysporum*, and constricted at the base. Spot tests negative. Not common, mostly on pine twigs, occasionally on other trees or on wood, on coast and in interior mountains statewide. Photo from a herbarium specimen collected in AB.

C. pinicola

Dimelaena
Moonglow lichens

Crustose lichens that have a rimose-areolate thallus with narrow, radiating lobes around the margin, generally without pruina. The black apothecia have thalline margins, and may be on the surface or immersed in the areoles; disks can be pruinose. Spores brown, 2-celled, thin, uniform walls, 8 per ascus. On rock.

Dimelaena californica
California moonglow lichen

Thallus dark brown, shiny when young, dull when older, usually with a black prothallus. Apothecia black, immersed. Most often found parasitic on

Dimelaena radiata or other crustose lichens, but when mature will grow on rock. Spot tests: K+ yellow to red, C–, KC–, P± orange. Uncommon, mostly coastal, central to southern CA. Compare to the more common *Buellia badia,* which it closely resembles, and to *D. thysanota,* which has a colorless hypothecium; *D. californica*'s hypothecium is dark brown. Photo from the Channel Is.

D. californica

Dimelaena oreina

Golden moonglow lichen

Thallus thin, pale greenish yellow, black between the areoles, and usually with elongated, conspicuously radiating marginal lobes. Apothecia lecanorine, either on the surface or without obvious margins and immersed in the areoles. Spot tests: seven chemical races have been distinguished, giving rise to a variety of reactions. Fairly common on rock, statewide, especially in mountains. The narrow lobes, its flatness and color, and its network of black lines make this species fairly easy to spot. Photo from southern ON.

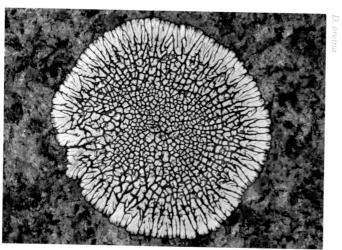

D. oreina

Dimelaena radiata
Silver moonglow lichen

Thallus variable in thickness, usually whitish to pale gray, and pruinose, but sometimes brown and without pruina. Black, often pruinose apothecia, typically with lecanorine margins; brown hypothecium. Spot tests negative. Fairly common on coastal rocks, occasionally inland, Mexican border to Sonoma Co.

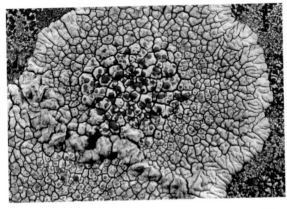

The gray form of *Dimelaena radiata*. Photo from coastal San Luis Obispo Co.

The brown form of *Dimelaena radiata*. Photo from inland San Luis Obispo Co.

Dimelaena thysanota
Mountain moonglow lichen

Thallus dark brown, without pruina or prothallus. Apothecia black, epruinose, mostly lecanorine, usually one in each areole. Hypothecium colorless. Spot tests negative. Moderately common in mountains statewide, more frequently inland than coastal. Compare with *D. californica*. A similar but rare species from central and southern Coast Range habitats is *D. weberi* (not pictured), with a yellowish thallus and areoles that become quite convex. Photo from a herbarium specimen collected in CA.

Diploicia canescens
Thallus lobed, bluish-white, forming rosettes 1–3 cm across, often pruinose on the lobe tips, with irregular clusters of soredia that are a bit darker than the

thallus. Apothecia lecideine, rare. Medulla white or yellowish. Looks rather foliose, like a species of *Physcia,* but lacks a lower cortex. Spores *Physcia*-type, 2-celled. Spot tests: cortex K+ yellow, KC−, C−, P− or P+ pale yellow; medulla, when white, no reactions; when yellow, K+ red, KC+ yellow, C+ yellow, P−. On wood, bark, or rock, mostly coastal, often near water, central to southern CA.

Diploschistes

Crater lichens

Crustose lichens with fairly thick, cracked to areolate thalli and apothecia conspicuously immersed in tiny "craters." Apothecia have a black or brown exciple and a thalline margin; usually they sit in an obvious cavity, but sometimes they can be almost closed, resembling perithecia. Without soredia or isidia. Spores muriform, dark brown, and sometimes shriveled-looking, 4–8 per ascus. On rock, soil, moss, or parasitic on other lichens. An additional CA species is *D. actinostomus* (not pictured), with thin, pale gray areoles and immersed apothecia resembling perithecia, found in the central and southern coast and mountains.

Diploschistes diacapsis

Desert crater lichen

Thallus white to dull gray, warty-areolate, pruinose. Apothecia concave, up to 2.5 mm wide, black, or gray from pruina, with margins. Spores 8 per ascus. Spot tests: K+ yellow, red, or purple, KC+ red or KC–, C+ dark red or C–, P–. On soil, occasionally calcareous rock, mostly in coastal locations, rarely inland, southern CA. Resembles *D. muscorum,* but that species is 4-spored and often parasitic. Photo from central TX.

D. diacapsis

Diploschistes muscorum

Cowpie lichen

Thallus and apothecia much like *D. scruposus,* except that *D. muscorum* is sometimes pruinose and consistently has 4 spores per ascus. It begins growing as a parasite on other species of lichens, especially *Cladonia,* and incorporates the algae from its host. As it matures, it becomes free-living on moss and soil, sometimes on wood. Rare on rock. Spot tests: K+ yellow becoming purple, KC+ red, C+ red, P–. Common and widespread, found almost everywhere that *Cladonia* species grow, coast and mountains. Compare with *D. diacapsis* and *D. scruposus*. Photo from coastal San Luis Obispo Co.

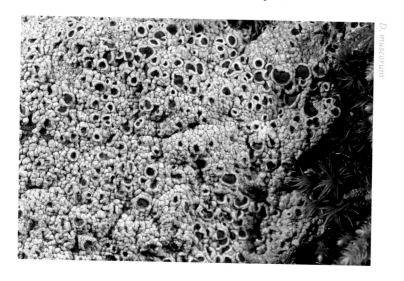

D. muscorum

Diploschistes scruposus
Crater lichen

Thallus areolate or verrucose, pale gray to almost white, epruinose. Apothecia black, slightly concave, with margins, sometimes double ones, embedded in craters, but these can vary from being almost closed to being open and broad, when they resemble apothecia of *Aspicilia* or *Lecanora*. Spores 4–8 per ascus. Spot tests: K–or K+ yellow to red or purple, KC+ red, C+ red, P–. On noncalcareous rock, in both coastal regions and mountains and coastal regions, statewide. This is the most common species of *Diploschistes*. Photo from the Gifford Pinchot National Forest, WA.

D. scruposus

Diplotomma
Button lichens

A crustose genus whose species used to be included in *Buellia,* but with 4-celled to muriform spores and sometimes thicker and more lobed thalli. Like *Buellia* species, the apothecia are black, lecideine. A few other species of *Diplotomma,* not described below, occur in CA, such as *D. alboatrum,* with a whitish to tan thallus, and black, often pruinose apothecia; spores 4-celled to submuriform with 5–8 cells; spot tests: cortex K– or K+ red; fairly common on rock and bark on the southern coast; and *D. venustum* (see *Buellia dispersa*).

Diplotomma penichrum
Poor button lichen

Thallus white to pale gray, continuous or with scattered areoles, sometimes with a black prothallus. Apothecia black with black margins. Spores submuriform to muriform, fewer than 12 cells per spore. Spot tests: thallus K+ yellow, C–, P+ yellow; medulla no reactions. On bark and wood of conifers, rare on deciduous trees, statewide, mostly in interior mountains, but a few collections from the Coast Range. Resembles *B. muriformis* (see *Buellia oidalea*). Photos from a herbarium specimen collected in WA.

D. penichrum

Dirina catalinariae
Catalina gray-disk lichen

Thallus thin, light to dark gray, with conspicuous granular soralia in rounded clumps (form *sorediata*) or more rarely with pale, lecanorine apothecia and no soredia (form *catalinariae*) or, very rarely, both apothecia and soralia. A scratch through the surface of the thallus will reveal the yellow color of the photobiont, the alga *Trentepohlia*. Spores colorless, fusiform, 4-celled, 8 per ascus. Spot tests: cortex and medulla K–, KC+ red, C+ red, P–. On coastal rock, especially vertical or overhanging rock faces, sometimes in extended colonies, central to southern CA. Another CA species, *D. paradoxa* (not pictured), is described under *Dendrographa conformis*.

The fertile form of *Dirina catalinariae*. Photo from San Simeon State Park, San Luis Obispo Co. Note the resemblance of this form to *Pertusaria californica*.

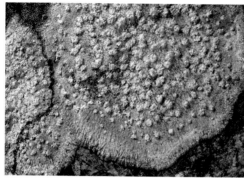

The sorediate form of *Dirina catalinariae*. Photo from the Channel Is.

Endocarpon pusillum
Stipplescale

Thallus small, squamulose, tan to brown or gray, with scales 0.5–3 mm across; forming irregular colonies on soil, rarely on rock. Black perithecia dot the upper surface. Lower surface brown to black with black, carbonized rhizines. Medulla white with a green algal layer. Hymenium with clusters of rounded

E. pusillum

cells of green algae. Spores brown, muriform with many cells, 2 per ascus. Spot tests negative. The photo shows a well-formed thallus, looking somewhat like a tiny *Dermatocarpon*, but *E. pusillum* most often grows in scattered patches of squamules that are visually hard to distinguish, and the appearance is quite variable. Resembles *Placidium*, but those lack hymenial algae and differ in the spores. Central and southern CA, at all elevations. A similar, less common species from the central and southern Coast Range, *E. loscosii* (not pictured), has a pale lower surface with sparse, long, pale rhizines. Photo from Zion National Park, UT.

Enterographa oregonensis

Leaf script

A tiny crust with a grayish to brown thallus. Apothecia rounded or slightly elongated and curved, tan with white margins. Spores fusiform and slightly curved, 6–8 celled, about 23–30 × 2.5–4.0 μm, with a gelatinous outer sheath. On evergreen leaves of huckleberry, and on conifer needles, central to northern coastal forests. The genus is mostly tropical, and species of *Enterographa* have not often been reported from CA, but this one has recently been found from Del Norte to Mendocino Cos. To the unaided eye it looks like a gray deposit on leaves or needles, not immediately recognizable as a lichen. Photo from a specimen collected in coastal Mendocino Co.

E. oregonensis

Fuscopannaria

Brown shingle lichens, Mouse lichens

Small, squamulose lichens, sometimes tightly appressed, brown, gray, or greenish, frequently with a blue to black prothallus around the margin. Apothecia brown to black, with a lecanorine or sometimes biatorine margin. Spores ellipsoid, colorless, 1-celled, 8 per ascus. Species in this genus used to be included within *Pannaria*. See also *Vahliella*, a genus segregated from *Fuscopannaria*.

Fuscopannaria coralloidea

Coral mouse lichen

Thallus small, brown to gray, squamulose, forming rosettes up to 4 cm wide, without an obvious prothallus. Apothecia numerous, reddish brown, without margins. The tiny lobes are coralloid, that is, with minutely branched cylindrical outgrowths. Spot tests negative. Occasional on soil, most often on road-cuts or riverbanks in moist coastal forests, statewide. The photo, from Six Rivers National Forest, shows the type specimen of the species.

F. coralloidea

Fuscopannaria leucostictoides

Petaled shingle lichen

Thallus flat, squamulose, up to 3 cm wide, with radiating lobes, or areoles less than 1 mm across and 2 mm long on top of a blue-black prothallus. Surface is olive-green but appears bluish gray because of pruina on the lobe tips. Apothecia brown, lecanorine, with grayish squamulose margins; without soredia or

F. leucostictoides

isidia. Spot tests: K+ yellow, C–, KC–, P–. Occasional on bark and twigs, usually of conifers, in coastal forest habitats, central to northern CA. Photo from Olympic National Park, WA.

Fuscopannaria praetermissa
Moss shingle lichen
Thallus olive to reddish brown, with thick squamules that ascend and overlap, forming a continuous layer over the substrate; lobes have pale gray felty tips, usually with fingerlike outgrowths resembling isidia. Apothecia, when present, brownish, with or without thalline margins. Spot tests negative. On moss, especially over calcareous soil, occasionally on burned stumps or decaying wood, rare on rock, in the Sierra Nevada, but in coastal areas as well. The name may actually represent several species, however. Resembles *Massalongia carnosa*, which has longer, strap-shaped lobes without felty tips and different spores. Photo from the Coast Range, OR.

F. praetermissa

Also of Note
Another CA species, not pictured, *Fuscopannaria cyanolepra*, forms squamulose patches on soil up to 10 cm wide; thallus is blue-gray to brownish with bluish marginal soredia and lacks a distinct prothallus; apothecia rare; spot tests negative; wet, could be mistaken for a sorediate species of *Collema*; Coast Range and inland mountains statewide.

Graphis scripta
Common script lichen
Crustose lichen with a thin, white to pale gray thallus that often grows within the substrate and is barely visible. Fruiting bodies are black lirellae, elongated ascomata with branches that look like calligraphy. Spot tests negative. On

bark, usually in the shade, in coastal areas, central to northern CA. The genus *Graphis* has many species in North America, most in the Southeast, especially Florida, but *G. scripta* is the only one usually encountered in CA. Can resemble bark-dwelling species of *Opegrapha,* which has a different spore type; some species of *Arthonia* also have lirellae-like fruiting bodies. When immature, the apothecia can be gray rather than black and immersed in irregularly shaped white areoles. Another species, uncommon in CA, is *G. elegans* (not pictured), with multilayered lirella walls; spot tests: K+ yellow turning reddish in the cortex and medulla, KC–, C–, P+ orange; on bark, notably on cypress cones, central coast.

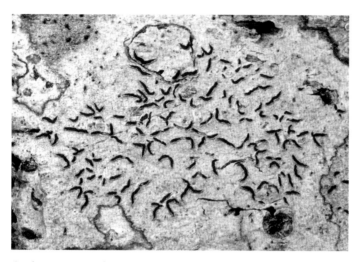

Graphis scripta in its classic, mature form. Photo from Humboldt Redwoods State Park.

The immature *Graphis scripta* can look rather different. Photo from Inverness, Marin Co.

Gyalecta jenensis

Rock dimple lichen

Thallus thin, gray to pale pink, orange or greenish. Apothecia yellow-pink with thick, pale, waxy margins that can be even or cracked and broken into segments. Spores colorless, 4–8 per ascus. Spot tests negative. Uncommon, on calcareous rocks or mortar in coastal areas, central to southern CA. A similar-looking and somewhat less rare species, *Gyalecta herrei* (not pictured), grows on the bark of half-dead trees, central to southern CA. Photo from a herbarium specimen collected in Ireland.

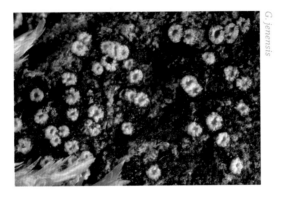

G. jenensis

Heppia conchiloba

Common soil ruby

Thallus squamulose to almost foliose, with lobes that are often concave and up to 8 mm wide, light brown but with pruina, giving it a gray appearance. Apothecia reddish brown, immersed, one or more per lobe, looking like concave spots. Lobe margins can be rough, almost granular. Spores ellipsoid, colorless, 1-celled, 8 per ascus. Spot tests negative. Occasional on soil in southern

H. conchiloba

interior deserts. Somewhat similar and more uncommon, and not pictured here, are *H. adglutinata* (olive-colored, epruinose), *H. despreauxii* (mottled, with reticulate depressions), and *H. lutosa* (black granular thallus, resembling *Collema*). Compare with *Peltula* and *Placidium*. Photo from Austin, TX.

Hydropunctaria maura
Sea tar, Black seaside lichen
A crustose lichen with perithecia, formerly in the genus *Verrucaria*. Thallus very dark brown to black, usually cracked into irregular areoles, with a black basal medullary layer; perithecia appear as rounded bumps. Spores colorless, 1-celled, 8 per ascus. Spot tests negative. Forms irregular patches on ocean-side rock in the upper tidal zone, northern coast to San Francisco Bay area. Compare with *Verrucaria* and *Staurothele*. Some free-living green algae in the upper tidal zone can look very much like *H. maura,* but they lack fungal hyphae or ascomata. Photo from Tomales Bay, Marin Co.

H. maura

Hypocenomyce
Clam lichens
Small squamulose lichens. The squamules are often scattered, but when clustered they can resemble a primary thallus of *Cladonia*. Apothecia biatorine, black or dark brown. Spores ellipsoid, colorless, 1-celled, 8 per ascus. On wood or bark, often on burned wood.

Hypocenomyce anthracophila
Small clam lichen
Squamules tiny, greenish to brown, up to 0.8 mm wide, overlapping or scattered, with fine gray soredia in labriform soralia on the underside of upturned squamules. Apothecia reddish brown, but often not present. Spot tests: upper cortex and soralia K–, KC–, C–, P+ orange/red; medulla K–, KC–or KC+ purple,

C–, P+ red, UV+ white. Occasional on conifer bark, especially when charred from fire, in the Sierra Nevada and interior mountains. A less common species, found in the Sierra Nevada and southern mountains, is *H. castaneocinerea* (not pictured); it has darker, more convex squamules with brown margins and brown soralia; spot tests: cortex P–. Photo from a herbarium specimen collected in MI.

Hypocenomyce scalaris
Common clam lichen

Squamules overlapping, olive or pale green to brownish, up to 2 mm wide, usually ascending, with greenish powdery soredia in labriform soralia on the underside and margin of each squamule. Apothecia uncommon, black;

sometimes with black pycnidia. Spot tests: K–, KC+ red, C+ pink or red, P–. Most often on charred wood, but also on conifer bark, statewide, coast and mountains. The most common species of *Hypocenomyce*. Similar but rarer is *H. sierrae* (not pictured), with smaller, greener, more crenulate squamules and smaller, epruinose apothecia; spot tests: K+ yellow, P+ yellow; in the Sierra Nevada and southern mountains, often with *H. scalaris*. Photo from Contra Costa Co.

Lecanactis californica
California old-wood lichen
Thallus crustose, variable, thin or thick, rimose to areolate or almost squamulose, pale gray to pinkish. Apothecia black but often gray-pruinose, with thin, black, lecideine margins that may disappear in older fruiting bodies. Spores cigar-shaped, colorless, 4-celled, 8 per ascus. Spot tests: K+ pale yellow to orange or K–, C+ pale yellow or C–, KC–, P+ yellow to orange. Occasional on bark or seashore rock, Monterey Co. to Mexico. Photo from San Simeon State Park, San Luis Obispo Co.

L. californica

Lecania
Crustose to squamulose lichens with thalli of variable thickness and colors from pale gray to dark brown, often yellowish or reddish. Apothecia lecanorine or biatorine, generally dark brown, often pruinose. Spores colorless, 2–4-celled. On bark and rock. Compare with species of *Lecanora*, which differ in ascus type and other microscopic features and are usually more conspicuous. Quite a few *Lecania* species, not described below, occur in CA. One of the most common is *L. cyrtella*, with a very thin, pale grayish thallus; apothecia yellow, red, brown, or black; spores narrowly elliopsoid, colorless, 1–4-celled, typically 9–14 × 4–5 µm; on bark, mostly in central and southern coastal areas.

Lecania brunonis

Brown rim lichen

Thallus squamulose to areolate, brown to gray when dry, greenish when damp. Apothecia brown with lecanorine margins that become discontinuous and thin with age; disks can become redder and paler when damp. Thallus often with pycnidia. Spores ellipsoid, colorless, 2-celled (sometimes 4-celled), typically 10–15 × 4–6 µm, 8 per ascus. Spot tests negative. On siliceous rock, occasionally on bone, mostly coastal, central to southern CA. The most common *Lecania* species in CA. A similar species on rock is *L. hassei* (not pictured), with longer, fusiform ascospores, 2-celled, typically 14–22 × 3.5–4.5 µm; coastal southern CA. Photo from the Channel Is.

L. brunonis

L. dudleyi

Lecania dudleyi
Dudley's rim lichen

Thallus coarse, bullate to squamulose, various shades of brown, gray, or yellowish, sometimes with a black prothallus and sometimes pruinose. Apothecia brown to black with thin, thalline, white to brown margins; often with immersed pycnidia. Spores broad, ellipsoid, colorless, 2-celled, typically 11–15 × 6–7 μm, 8 per ascus. Spot tests negative. On sedimentary rock or soil near the coast, Monterey Co. to Mexico. Photo from the Channel Is.

Lecania fructigena
Fruiting rim lichen

Thallus variable from papillate to more or less areolate or subsquamulose, thin or thick, with elevated, small, round verrucae (warts); pale yellowish to gray or brown. Apothecia sessile, becoming almost stipitate, brown to gray with thin, pale rims, usually epruinose but most often with numerous pycnidia. Spores oblong to ellipsoid, colorless, 2-celled, typically 10–15 × 4–5 μm, 8 per ascus. Spot tests: K+ yellowish, KC–, C–, P–. On rock in coastal, often maritime, habitats, mostly southern to central CA but as far north as Mendocino Co. Photo from San Luis Obispo Co.

L. fructigena

Lecania pacifica
Pacific rim lichen

Thallus rimose to areolate, thin, minutely lobed at the margins, brown to gray or greenish, often with a darker prothallus. Apothecia sessile, brown, with paler margins. Spores oblong-ellipsoid, colorless, usually 2-celled, typically 12–18 × 4–5 μm, 8 per ascus. Spot tests negative. Occasional on acidic rock in coastal habitats, central and southern CA. Endemic to CA. Photo from Montano de Oro State Park, San Luis Obispo Co.

Lecanographa

The genus is similar to *Opegrapha,* but the ascus type and spores are different, and the ascomata are usually pruinose.

Lecanographa brattiae
Lithic scribble lichen

Thallus rimose to areolate, creamy white or grayish, often with a black prothallus; numerous black lirellae, sometimes branched, making it look like a species of *Opegrapha.* Spores 3–4-celled, without a visible gelatinous sheath. Spot tests: thallus K+ yellowish, C+ reddish, P–. Occasional on coastal rock, Monterey Co. to Mexico. Photo from the Channel Is.

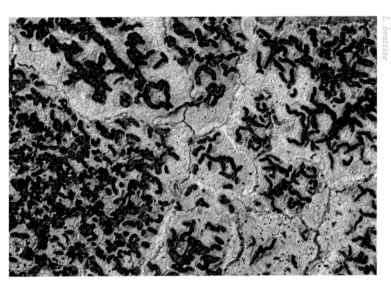

Lecanographa hypothallina
California chalk-crust

Thallus usually thick and cushionlike, chalky white, rimose-areolate to verrucose, heavily covered in white pruina. Apothecia extremely variable, usually rounded, convex and sitting above the surface with pale margins, but sometimes elongate, sunken, and with black margins, or intermediate with irregular apothecia lacking margins. Spores oblong to fusiform, colorless, becoming brown when old, 8-celled, 8 per ascus, with a conspicuous gelatinous sheath. Spot tests: thallus K–, KC+ red, C+ red, P– or P+ pale yellow. On coastal, usually shaded, rock, southern CA to Monterey Co.

A common form of *Lecanographa hypothallina*. Photo from San Luis Obispo Co.

The scriptlike form of *Lecanographa hypothallina*. Photo from the Channel Is.

Lecanora
Rim lichens

A large and diverse genus of mostly crustose lichens; a few species are fruticose. Variable in thallus type and color, but apothecia usually with "lecanorine" margins consisting of thallus-like tissue. The genus is primarily defined by its crustose habit, ascus type, and 1-celled, colorless spores that are spherical to ellipsoid, usually 8 per ascus. Chemistry is varied, and several subgroups are distinguished, such as the *Lecanora subfusca* group, with brown apothecia, a gray K+ yellow thallus, and crystals in the apothecial margins, and the *Lecanora albella* group, with pruinose apothecia, an amphithecium with oxalate crystals, and with thallus and apothecial margins that react K+ yellow owing to the presence of atranorin. Most crustose lichens one sees that have conspicuous round apothecia with contrasting margins and are not bright orange like *Caloplaca* are likely to be species of *Lecanora*, although they can be confused with species of *Ochrolechia*, which have orange-tan apothecia on a white thallus, or *Lecania*, *Protoparmelia*, *Rhizoplaca*, *Rinodina*, and *Tephromela*. On all substrates. Many other species of *Lecanora*, not described below, occur in CA.

Lecanora albellula
Wood-spot rim lichen
Thallus thin or absent, with yellowish gray to tan dispersed warts or areoles. Apothecia less than 0.5 mm wide, ocher to black with thalline margins, sometimes with thin pruina. Epihymenium with fine granules soluble in K. Spores narrowly ellipsoid. Spot tests negative. Occasional on wood and bark in interior mountains, statewide. Similar to *L. subintricata* (not pictured), which has biatorine apothecial margins and is rare in CA. Specimens with dark apothecia can resemble *L. saligna,* but that species has apothecia often larger than 0.5 mm and ellipsoid to broadly ellipsoid spores. Photo from the Rocky Mountains, CO.

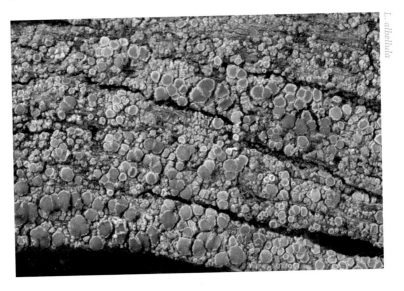

L. albellula

Lecanora allophana
Brown-eyed rim lichen
Thallus white to pale gray, continuous or verrucose-areolate, smooth to rough, epruinose but sometimes with granular soredia. Apothecia large, up to 2 mm, reddish brown to dark brown, constricted at the base, margins pale. Amphithecium with small crystals that extend from the medulla into the cortex. Epihymenium red-brown, without granules. Spot tests: K+ yellow, KC–, C–, P– or pale yellow. Occasional on bark, mostly in the Coast Range, statewide.

L. horiza (not pictured) is similar but has smaller apothecia and the amphithecial crystals are confined to the medulla; fairly common on bark in coastal southern CA. Photo from Mt. San Jacinto, Riverside Co.; note the distinct color variation in the apothecia.

L. allophana

Lecanora argopholis

Varying rim lichen

Thallus variable, gray to pale yellow or yellow-green, verrucose-areolate to distinctly lobed or even building into more or less cylindrical verrucae and appearing subfruticose. Apothecia red-brown to black, 0.5–3 mm wide, with prominent margins. Amphithecium with small crystals (insoluble in K). Epihymenium red-brown, without crystals. Spot tests: cortex K+ yellow, C–, KC+

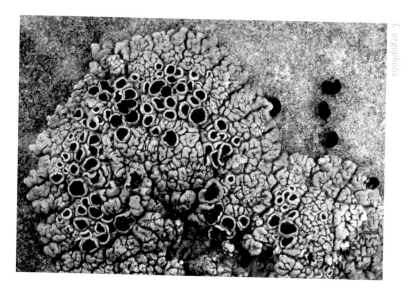

L. argopholis

yellow, P– or P+ pale yellow; medulla usually K– but sometimes pale yellow, C–, KC–, P–. A species of the Rocky Mountains and Great Basin. Uncommon in CA, usually on rock, occasionally on moss or plant detritus, mostly south and east of the Sierra Nevada crest. Can resemble *L. cenisia*, but that species is usually pruinose. Photo from UT.

Lecanora caesiorubella
Frosted rim lichen

Thallus pale gray, smooth to areolate or verruculose. Apothecia fairly large, up to 3 mm wide, sessile, round to irregular, pale gray but often pinkish or lavender due to a coating of pruina; margins usually thick and flexuose. Amphithecial medulla with small crystals that dissolve in K. Spot tests: apothecial sections K+ yellow, or yellow turning red, KC–, C–, P+ red or (rarely) P–. *L. caesiorubella* has a number of chemical variants in North America, some of which have been termed subspecies though their taxonomic status is uncertain. It is one of the most common bark-dwelling *Lecanora* species in CA. A similar species, occurring on bark in the central and southern coast and mountains, is *L. albella* (not pictured), with small, pruinose apothecia that have thin margins. Compare also with *L. carpinea*. Photo from San Luis Obispo Co.

Lecanora californica
California rim lichen

Thallus areolate to more or less verrucose, white to yellowish gray. Apothecia large, 0.9–2.5 mm wide, round or irregular and often crowded together, disks brown to black, constricted at the base, often pruinose, with thick, white margins that are usually wavy. Amphithecium with large crystals (insoluble in K). Epihymenium with small crystals (soluble in K). Spot tests: K+ yellow, KC–, C–, P– or pale yellow. Fairly common on rock in coastal parts of the San Francisco Bay region and Sonoma Co., and on the Channel Is., occasional elsewhere. Resembles *L. cenisia*, but usually more pruinose and with more crowded apothecia that are conspicuously constricted at the base, unlike *L. cenisia*, which has closely attached apothecia. Photo from the Channel Is.

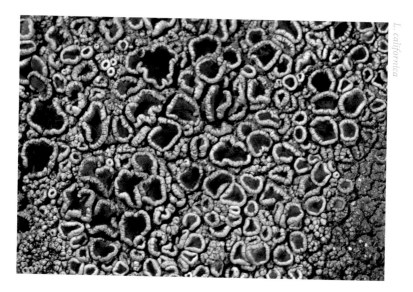

L. californica

Lecanora campestris
Field rim lichen

Thallus continuous to verrucose-areolate, pale gray to yellowish. Apothecia reddish brown, sessile, 0.5–1.6 mm wide, with thin, even, pale margins. Amphithecium with a distinct cortex and small crystals that are insoluble in K. Epihymenium red-brown, without crystals, but with pigment dissolving in K. Spot tests: K+ yellow, KC–, C–, P– or pale yellow. Usually on soil or calcareous rock, sometimes on bone, in southern coastal areas north to at least Marin Co. Photo from Marin Co., on an old whale skull.

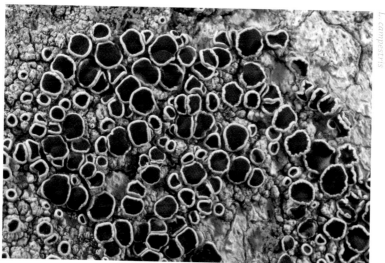

L. campestris

Lecanora carpinea

Hornbeam rim lichen

Thallus continuous or rimose-areolate, pale gray or greenish white. Apothecia sessile, 0.5–1.3 mm wide, pale tan to pale pink but covered with pruina that give them a gray to bluish tinge. Apothecial margins prominent, the same color as the thallus, with a well-developed amphithecial cortex. Amphithecium and epihymenium with small crystals that dissolve in K. Spot tests: thallus K+ yellow, apothecial disk C+ orange to reddish, thallus and apothecial sections P– or pale yellow. Occasional on bark and wood, Coast Range and inland mountains, statewide. Photo from the Rocky Mountains, CO.

L. carpinea

Lecanora cenisia

Smoky rim lichen

Thallus almost white, or pale yellowish gray, usually thick and verrucose. Apothecia adnate or sessile, sometimes constricted at the base, 0.5–2 mm wide, various shades of tan in sheltered locations or almost black where exposed, usually with a coating of light pruina. Apothecial margins thallus-colored, thick, smooth. Amphithecium with large crystals (insoluble in K). Epihymenium granulose with small crystals (soluble in K). Spot tests: thallus K+ yellow, C–, KC–, P– or pale yellow. On rock or hard, weathered wood in inland mountains and the Coast Range, statewide. Rather variable in appearance, and one of the most common species of *Lecanora*, part of the *subfusca* group. Compare with *L. californica,* which has larger, more pruinose apothecia, and is strictly coastal, and *L. campestris,* which has no pruina and therefore lacks granules in the epihymenium, and has smaller amphithecial crystals. Photo from eastern side of the Cascades, OR.

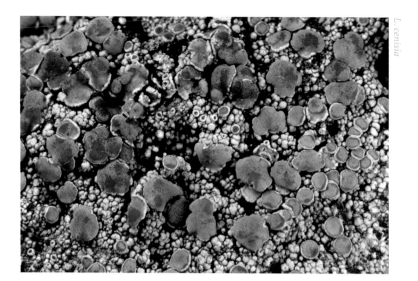

Lecanora chlarotera

Bark rim lichen

Thallus pale gray or slightly yellowish or greenish, continuous or rimose-areolate to verrucose-areolate, without pruina. Apothecia sessile, reddish or orange-tan with pale, smooth or rough margins. Amphithecium with large crystals (insoluble in K). Epihymenium granulose with a layer of coarse granules on top and small crystals that dissolve in K. Spot tests: thallus K+ yellow, C–, KC–, P– or P+ yellow or orange to red. On deciduous bark in mountains, statewide in inland ranges. Resembles *L. hybocarpa* in appearance but differs in the type of epihymenial crystals; microscopic examination is necessary to tell them apart. Photo from Carson National Forest, NM.

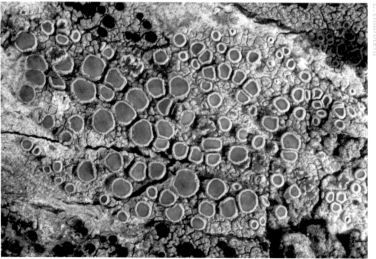

Lecanora cinereofusca
Beaded rim lichen

Thallus pale gray to nearly white, smooth or cracked to verruculose. Apothecia orange to reddish brown, epruinose, at first immersed and level with the thallus, then emerging with a broken, beaded-looking margin. Amphithecium with large crystals. Spot tests: epihymenium and often apothecial margin P+ orange. Occasional on bark, usually of deciduous trees, northern Coast Range. Photo from north coastal WA.

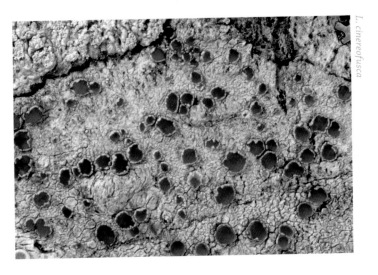

L. cinereofusca

Lecanora circumborealis
Black-eyed rim lichen

Thallus thin, usually continuous but sometimes areolate, pale gray to white, frequently with a black prothallus. Apothecia dull, very dark brown to black, 0.4–0.8 mm wide with conspicuous margins, epruinose. Amphithecium with thick cortex, broad at the base, and with few to many large crystals (insoluble

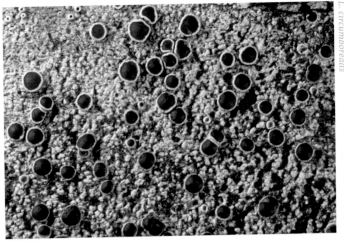

L. circumborealis

in K). Epihymenium embedded with tiny brown granules between the tips of paraphyses. Spores 13–17.5 × 8–11 µm, thick-walled. Spot tests: apothecial section K+ yellow, KC−, C−, P− or pale yellow. On bark, Coast Range and inland mountains, statewide. *Lecanora hybocarpa* has a more uniformly thickened cortex. A similar but less common species on bark or wood is *L. pulicaris* (not pictured), with paler apothecia, smaller spores, and a smaller amphithecial cortex thinner at the base than in *L. circumborealis;* spot tests: medulla P+ red; occasional statewide. Photo from southern ON.

Lecanora confusa
Bumpy rim lichen
Thallus of granular warts that can coalesce into a rimose-areolate crust, pale tan to gray or yellowish, sometimes with globose soredia. Apothecia 0.45–0.9 mm, usually sessile with a constricted base, or rarely appressed, orange-tan, often crowded. Spores narrowly ellipsoid. Spot tests: thallus and amphithecium K−, C+ orange, KC+ orange, P−. On bark and wood in coastal areas, Mexico to Marin Co. Resembles *L. strobilina* and *L. symmicta,* but these have a C− thallus; *L. symmicta* (not pictured), has biatorine margins and yellowish apothecia. Photo from a driftwood log, coastal San Luis Obispo Co.

L. confusa

Lecanora demosthenesii
Marbles rim lichen
Thallus yellowish to gray or tan, verruculose to areolate. Apothecia tan, constricted at the base, epruinose or slightly pruinose. Amphithecium with large crystals insoluble in K. Epihymenium clear to yellow-brown, with pigment and crystals that dissolve in K. Spores ellipsoid, fairly narrow. Spot tests: K+ yellow, KC−, C−, P+ pale yellow. Occasional on bark or wood, rarely on rock, central and southern coast. Can resemble pale forms of the more common *L. pacifica,* but that species has broader spores. Photo from Pt. Lobos, Monterey Co.

Lecanora dispersa

Mortar rim lichen

Thallus generally not visible, growing between crystals of the rock substrate. Apothecia brownish, clustered or dispersed, with white margins, epruinose. Epihymenium with small granules. Spot tests usually negative. Typically on calcareous rock or concrete but also on siliceous rock, bark, or even other lichens, coast and inland, statewide. It has a higher tolerance for pollution than most lichens and is often seen on concrete and mortar in urban areas. Similar in appearance but less common is *L. crenulata* (not pictured), which has cracked apothecial margins and is sometimes pruinose; central and southern coastal areas. Also resembles *L. hagenii*, but that species has pruinose apothecia and grows on bark. Photo from south rim of the Grand Canyon, AZ.

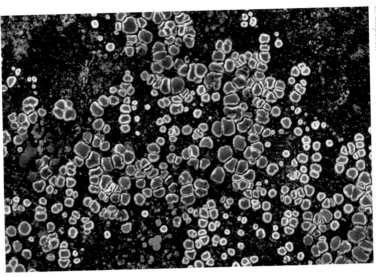

Lecanora gangaleoides
Knotted rim lichen

Thallus various shades of gray to yellowish, rimose, areolate or verrucose, epruinose. Apothecia usually black, sometimes dark brown, 0.7–1.2 mm wide, immersed when young, on the surface when mature, with gray to white rims. Amphithecium with large crystals insoluble in K. Epihymenium dark brown to greenish, without crystals. Spot tests: thallus and apothecial margin K+ yellow, KC–, C–, P+ pale orange. On rock, mostly along the coast but occasionally inland, Marin Co. to Mexico. Compare with *Tephromela atra,* which usually has a whiter thallus and always has a purplish hymenium. Photo from the Channel Is.

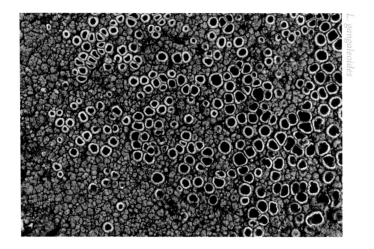

L. gangaleoides

Lecanora garovaglii
Sagebrush rim lichen

Thallus yellow-green, gray-green when pruinose, forming rosettes with sinuous, folded, convex lobes around the edge that sometimes blacken at the tips.

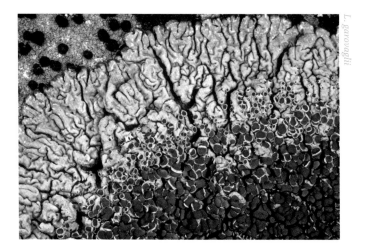

L. garovaglii

Apothecia pale to dark yellowish brown, with light margins. Amphithecium with colorless granules that are insoluble in K. Epihymenium orange to brown with small granules that dissolve in K. Spot tests: cortex K– or occasionally pale yellow, KC+ gold; medulla K–, C–, KC–, P– or (rarely) yellow. A Great Basin and desert species. On rock, especially sandstone, mostly in inland areas and the Sierra Nevada; rarely coastal. Somewhat resembles *L. muralis* but has thicker, more convex lobes. Photo from UT.

Lecanora hagenii
Hagen's rim lichen

Thallus thin and gray or entirely within the substrate. Apothecia 0.3–0.9 mm wide, greenish to brown, and pruinose, with gray to white, smooth to bumpy or dentate margins. Amphithecium and epihymenium with granules (insoluble in K). Spot tests negative. Usually on bark, wood, or other organic substrates, occasionally on rock; probably statewide but known mostly from central to southern CA, most often in coastal sites. Compare with *L. dispersa*, which has epruinose apothecia and grows on rock. A similar species, *L. crenulata* (not pictured), has larger, heavily pruinose apothecia with cracked margins and a thicker amphithecial cortex; on calcareous rock in the central and southern Coast Range. Photo from a herbarium specimen on rock collected in the Santa Monica Mountains, Los Angeles Co. Usually the species is more pruinose than the one in the photo.

L. hagenii

Lecanora hybocarpa
Bumpy rim lichen

Thallus pale gray, usually verruculose but sometimes smooth. Apothecia dull
orange to reddish brown, 0.4–1 mm wide, usually with bumpy margins, epruinose. Disk color is quite variable, even in the same location. Amphithecium
with large crystals insoluble in K. Epihymenium with tiny granules between
the tips of the paraphyses that do not dissolve in nitric acid. Spot tests: cortex
and apothecial sections K+ yellow, C–, KC–, P–. On hardwood bark, rarely on
conifers, in the Coast Range and southern inland mountains, statewide. For
similar species see comments under *L. circumborealis*. Photo from coastal MA.

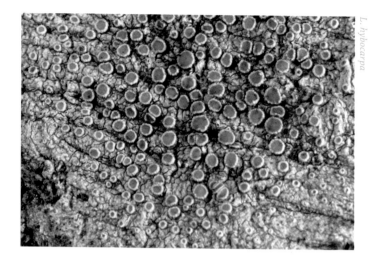

L. hybocarpa

Lecanora mellea
Honey-colored rim lichen

Thallus distinctly brown, forming lobed rosettes. Apothecia orange-brown,
sometimes sparse. Upper cortex with dead algal cells. Amphithecium without

L. mellea

granules. Epihymenium with fine granules (soluble in K). Spot tests negative. On siliceous rock, commonly granite, mostly in inland mountain ranges, especially the Sierra Nevada. Similar to *L. pseudomellea* (not pictured), which is often pruinose, has a true cortex, and sometimes reacts P+ orange or yellow in the medulla; fairly common in the Sierra Nevada. Photo from the eastern side of the Sierra Nevada.

Lecanora muralis
Stonewall rim lichen
Thallus pale yellowish green, somewhat waxy looking, usually lacking pruina, often forming lobed rosettes but can be irregular in shape. Apothecia yellowish tan, usually clustered toward the center of the thallus. Amphithecium without granules or only a few, or a few large crystals. Epihymenium with fine granules, soluble in K. Spot tests: cortex KC+ gold, medulla all tests negative. On rock, occasionally on soil, very rarely on bark, statewide, in all habitats. Probably the most widespread and common rock-dwelling species of *Lecanora* in CA. Paler and greener than *L. mellea* and yellower than *L. sierrae*. Quite variable; coastal thalli can be very pale, almost whitish. Photo from Mt. Hamilton, Santa Clara Co.

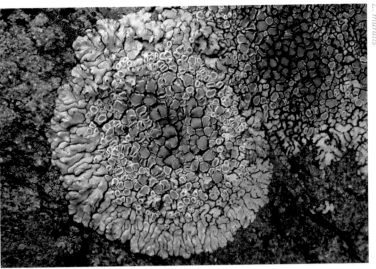

L. muralis

Lecanora pacifica
Multicolored rim lichen
Thallus thin to bumpy, yellowish gray, often with a blue-black prothallus. Apothecia large, flat, 0.7–1.2 mm wide, yellowish tan to black, even on the same disk, with white margins that are usually smooth, often with white pruina. Amphithecium with large crystals (insoluble in K). Epihymenium reddish brown with coarse granules on the surface and pigment and crystals that dissolve in K. Spores large, broadly ellipsoid. Spot tests: apothecial sections K+ yellow, C–, KC–, P–. Usually on bark of deciduous trees, occasionally on

conifers, statewide, especially in coastal habitats, but occasionally in the Sierra Nevada and northern interior mountains. One of the most common species of *Lecanora* in CA. Quite variable, so may be confused with other species; spore size and chemistry will distinguish it; compare pale apothecial forms with *L. confusa*. Photo from coastal WA.

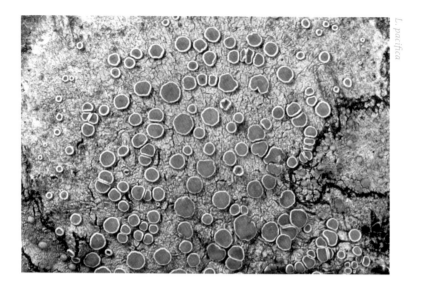

Lecanora phryganitis

Shrubby rim lichen

Thallus small, fruticose, forming pale yellow-green to tan cushions of short rounded stems that are very rough and granular and have granular soredia at the tips. Apothecia irregularly round and tan, with rough margins. Spot tests: cortex K+ yellowish, KC+ red-orange, C+ orange, P–. Quite distinctive,

although uncommon, on coastal rocks in the fog zone, Marin Co. to Monterey Co. Endemic to CA. Photo from Marin Co.

Lecanora pinguis
Seaside sulphur-rim lichen
Thallus thick, lumpy, pale yellow-green, often somewhat lobate at the edges, and frequently with a scabrous, rough surface. Apothecia closely attached, somewhat irregular, yellowish tan. Spot tests: cortex K+ yellow, KC+ red orange, C+ deep yellow, P–; medulla similar except KC+ orange. Often grows in the same locations as *L. phryganitis*, but more common and with a wider range. On coastal rocks in the fog zone, Sonoma Co. to Mexico. Photo from Marin Co.

L. pinguis

Lecanora polytropa
Granite-speck rim lichen
Thallus absent, or present only as scattered pale yellow granules or areoles. Apothecia flat, waxy, pale orange to yellow, 0.3–1.3 mm wide, in clusters or scattered, epruinose, with margins paler than the disk. Spores broadly ellipsoid. Epihymenium ocher, with granular crystals that dissolve in K. Spot tests: K–, KC+ yellow, C–, P–. Fairly common, on siliceous exposed rock, often granite, rarely on wood, in the Sierra Nevada and interior mountains. An extremely variable species. Specimens without a conspicuous thallus look identical to *L. stenotropa* (not pictured), which differs in having narrowly ellipsoid spores; occasional in the southern mountains. The photo, from southern ON, shows a specimen with a thallus that is more visible than usual for the species.

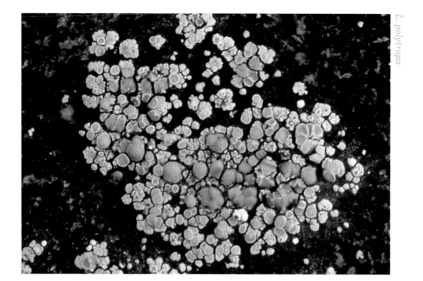

Lecanora rupicola
White rim lichen

Thallus rimose-areolate, usually thick, yellowish white or slightly greenish. Apothecia often immersed in the thallus when young but becoming sessile, sometimes slightly convex, dark brown to black, 0.5–2 mm wide, but covered with a thick, creamy white to pale blue pruina. Amphithecium and epihymenium with crystals that dissolve in K. Spot tests: cortex and medulla K+ yellow, KC–, C–, P–. Occasional on rock in mountains, statewide. It is similar to the much rarer *L. bicincta* (not pictured), which has a yellowish gray, epruinose thallus, and apothecia with a conspicuous blue-black ring around the disk just inside the thalline margin. Photo from southern ON.

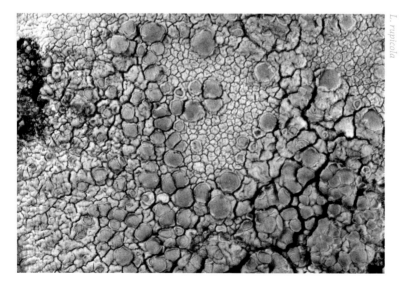

Lecanora saligna
Spotty rim lichen

Thallus thin, poorly developed, yellowish gray, with dispersed areoles. Apothecia orange-brown, sometimes crowded, with yellowish margins usually lighter than the disk. Amphithecium and epihymenium with granules that dissolve in K. Spot tests negative. Occasional on wood or conifer bark in the Sierra Nevada and southern interior mountains, rarely along the coast. Compare also with *L. albellula*. Photo from a herbarium specimen collected in BC.

L. saligna

Lecanora sierrae
Sierra rim lichen

Thallus areolate but lobate at the edges, usually forming tightly adnate, yellow-green rosettes (paler and grayer in the shade). Apothecia orange to brown, sometimes shiny, but duller in shade. Epihymenium with fine granules that dissolve in K. Spot tests: thallus K–, C–, KC+ yellow, P–; medulla KC–, P– or P+ orange or yellow, depending on chemotype. Resembles *L. muralis,* but usually a darker and more bluish green, with more convex lobes, and different in microscopic features. Also resembles *L. pseudomellea,* but that species has a yellowish to reddish brown thallus (see comments under *L. mellea*). Photo from the Mojave Desert mountains.

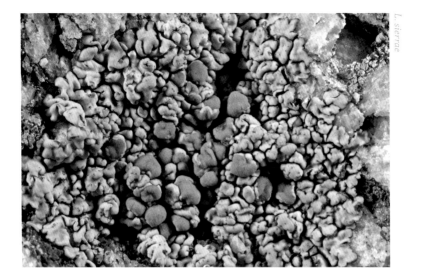

Lecanora strobilina

Mealy rim lichen

Thallus thin, cracked to granular, pale yellow-green to greenish gray. Apothecia small, 0.3–0.9 mm wide, waxy pale yellow, or slightly pinkish, usually crowded, with granular or sorediate margins. Amphithecium and epihymenium with granules that dissolve in K. Spot tests: K–, KC+ yellow or KC–, C–, P–. On bark, especially pines, and wood in well-lit locations, mostly coastal but very occasionally inland, central to southern CA. Resembles *L. confusa*. Similar is *L. substrobilina* (not pictured), with apothecial margins that are granular and knobby, narrower spores, and apothecial margins with entangled, gelatinized hyphae covering the algal layer; spot tests: C+ red; fairly common on bark in coastal habitats statewide. Photo from VA.

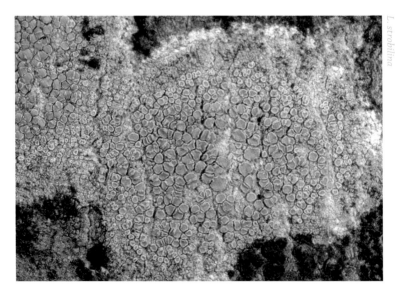

Also of Note

Of the CA species of *Lecanora* not described above, a few of the more common ones include:

Lecanora cadubriae, with a yellowish, granular to verrucose-areolate thallus; apothecia dark brown to blackish, 0.2–0.7 mm wide; asci more like *Biatora* than *Lecanora;* spot tests: K+ yellow to red, C–, KC–, P+ orange; on bark in mountains, statewide.

Lecanora expallens, yellow-green, often with a white to bluish prothallus; with soralia bearing farinose soredia that finally coalesce and become almost leprose; apothecia occasional, sparse, yellow to pink or brown, up to 1.5 mm wide; spot tests: K+ pale yellow, C+ deep yellow to orange, KC+ orange, P–, UV+ dull violet (short-wave), dull brownish (long-wave); on bark, usually of shrubs, statewide and strictly coastal.

Lecanora pseudistera, with a yellowish gray, areolate to verrucose or almost squamulose thallus, epruinose; apothecia red-brown, 0.4–1.5 mm wide, epruinose; amphithecium with large crystals dissolving in K; epihymenium without crystals; spot tests: thallus and apothecial margin K+ yellow, C–, KC–, P– or P+ pale orange; on rock, central and southern Coast Range and occasionally in the Sierra Nevada.

Lecanora semitensis, with a thin, yellowish to green, areolate thallus, sometimes pruinose, areoles scattered or contiguous; apothecia various shades of brown and often thinly pruinose, with margins lighter than the thallus, 0.5–1.5 mm wide; spot tests: thallus and apothecia K–, C–, KC–, P–; cortex KC+ yellow, blackened parts KC+ green, N+ red-violet; medulla KC–; fairly common, on acidic rock, typically on steep or shaded surfaces, in mountains statewide, mostly, but not always, inland.

Lecanora subimmergens, with a thin, grayish, rimose-areolate thallus, epruinose, sometimes with a white prothallus; apothecia red-brown, 0.4–1.4 mm wide, epruinose; amphithecium with large crystals insoluble in K; spot tests: thallus and apothecial margin K+ yellow, C–, KC–, P+ yellow; on rock along the central and southern coast and in southern mountains.

Lecidea

Disk lichens, Tile lichens

Crustose lichens with black lecideine apothecia, superficial or sunken, that have black margins. Thalli can be pale gray to brown or rusty orange, thick to entirely within the rock substrate and invisible, rimose to areolate, squamulose or almost lobate. Asci mostly K/I– but with a thin K/I+ blue cap at the tip. Spores colorless, 1-celled, 8 per ascus. The genus is still undergoing major taxonomic adjustments; in the narrowest concept, *Lecidea* species are found only on rock. Some, particularly those in the *L. atrobrunnea* group, are common and widespread. Some brown species resemble *Rhizocarpon bolanderi,* a common species in the Sierra Nevada. A number of other species, not described below, occur in CA.

Lecidea atrobrunnea

Brown tile lichen

Thallus extremely variable, but most often shiny brown, areolate to almost squamulose, usually with a black prothallus around the edge. Apothecia flat or somewhat convex, black, with thin, raised, black margins. Spot tests negative except medulla IKI+ blue-black. Several subspecies have been distinguished, mainly on the basis of chemistry. The photos illustrate some of the variation in this species. Can sometimes resemble *L. fuscoatra;* also compare with *Miriquidica scotopholis. L. atrobrunnea,* in the broad sense, is one of the more common crustose, lecideine lichens on rock in montane CA.

The classic, and probably most common, form and color of *Lecidea atrobrunnea.* Photo from the Rocky Mountains, CO.

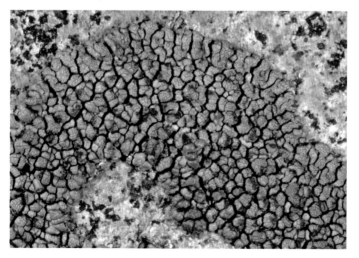

One of the variations of *Lecidea atrobrunnea* in form and color. Photo from the alpine Sierra Nevada, Inyo Co.

Lecidea brodoana

Brodo's tile lichen

Thallus squamulose, with the tiny lobes each having a pale gray margin; at first glance it looks more like a species of *Psora* than *Lecidea*. Apothecia black, with black, sometimes inconspicuous margins. Spot tests, including IKI, negative. The photo shows the holotype of the species, found on rock in the lower montane zone, Butte Co.; its true range is unknown, but it is listed as endemic to CA.

L. brodoana

Lecidea fuscoatra

Dusky tile lichen

Thallus brown, areolate, prothallus usually not showing. Apothecia flat or slightly convex, black, visible between the areoles. Spot tests: cortex K–, C+

L. fuscoatra

red, KC+ red, P−; medulla C−. Widespread and fairly common on rock, especially granite, statewide, especially in mountains but can be coastal as well. Rather variable and can resemble one of the forms of *L. atrobrunnea,* but *L. fuscoatra* has a dark hypothecium, has large spores, and usually reacts C+ red. A similar-looking species is *L. mannii* (not pictured), with an areolate, pale ocher to tan thallus and sometimes pruinose apothecia; fairly common, on acidic rock, southern coastal regions, especially mountains. It differs from *L. fuscoatra* in having an unpigmented hypothecium. Photo from Plumas National Forest.

Lecidea laboriosa
Working tile lichen
Thallus endolithic, usually appearing only as black apothecia on rock. Apothecia often lack visible margins or frequently have margins that are vertically striated in gray. Spot tests negative. Common and widespread, probably the most common endolithic species of *Lecidea* in CA. From southern CA north to at least Marin Co. or farther, both along the coast and in mountains. Compare with *Catillaria lenticularis, Polysporina simplex,* and *Sarcogyne clavus.* Photo from San Luis Obispo Co.

L. laboriosa

Lecidea plana
Flat tile lichen
Thallus fairly thick, pale gray, rimose-areolate, becoming dispersed around the edges. Apothecia black, slightly constricted at the base, with thin, raised margins. Spot tests negative. Uncommon, on rock, statewide. Photo from southern ON.

Lecidea tessellata

Tile lichen

Thallus chalky white to bluish gray, areolate, with a black prothallus around the outer edge; without soredia. Apothecia black, usually sunken between the tile-like areoles. Often forming large patches. Spot tests negative except thallus and medulla react IKI+ blue. Widespread and common, mostly in mountains, statewide.

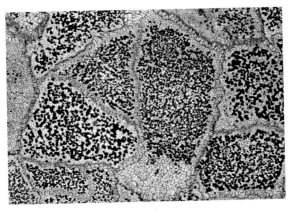

A distant view of *Lecidea tessellata*, showing its tilelike pattern. Photo from Acadia National Park, ME.

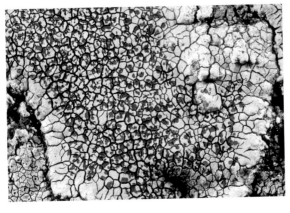

A closer view of *Lecidea tessellata*. Photo from San Bernardino National Forest.

Lecidella

Disk lichens

Crustose lichens, most with a gray or creamy thallus but some with little visible thallus; apothecia black, lecideine, usually shiny. Can resemble *Buellia*, *Lecidea*, or *Mycoblastus sanguinarius*, but species of *Lecidella* have paraphyses that separate easily in water or KOH, greenish tissues, broad, ellipsoid spores, and ascus tips that stain IKI+ dark blue. A few other species, not described below, occur in CA.

Lecidella asema

Creamy disk lichen

Thallus often lumpy, or rimose to areolate, yellowish or pale greenish. Apothecia black, lecideine, 0.5–1.5 mm wide. Spot tests: thallus surface K– or rarely K+ yellow, C+ orange, KC+ orange, P– or rarely P+ yellow. On siliceous rock, occasionally on wood, bark, or soil, mostly in coastal areas but also in southern mountains, Sonoma Co. to Mexico. The species formerly called *L. eleochromoides* is now considered to be *L. asema*. Photo from coastal San Luis Obispo Co.

L. asema

Lecidella carpathica

Carpathian disk lichen

Thallus areolate, verruculose or bullate, chalky white to pale yellowish gray, sometimes with pycnidia. Apothecia black, flat to convex, with thin black margins, epruinose. Spot tests: thallus surface K+ or rarely K–, C–, KC–, P+ yellow or rarely P–. Mostly on rock, occasionally on wood or bark, statewide. Compare with species of *Buellia*. Photo from San Luis Obispo Co.

L. carpathica

Lecidella euphorea

Lumpy disk lichen

Granular thallus, or rimose to areolate, pale yellowish or greenish gray. Apothecia black, epruinose, 0.8–1.2 mm wide, becoming convex with age. Epihymenium without crystals. Spot tests: thallus surface K+ yellow, KC+ yellow, C+ orange (but reaction can be poor), P+ yellow. Grows on wood or bark, especially at high elevations in both the Coast Range and the Sierra. Almost identical in appearance is *L. elaeochroma* (not pictured), which has tiny crystals on the surface of the epihymenium; on wood and bark, mostly in the Coast Range but also in the Sierra Nevada. Photo of a "compare with" specimen from Carson National Forest, NM.

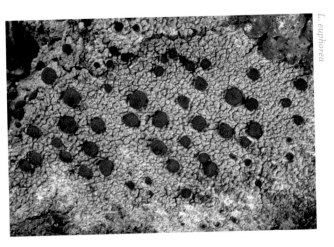

L. euphorea

Lecidella stigmatea

Gray disk lichen

Thallus variable, but usually thin, or growing within the rock, rimose-areolate to verruculose, pale gray to yellowish or greenish. Apothecia black to sometimes brownish, flat or convex. Spot tests: thallus surface K+ yellow, other reactions usually negative. On calcareous or noncalcareous rock, statewide, in mountains and along the coast. The colorless hypothecium of this species sets it apart from the other species of *Lecidella* illustrated here, all of which have a yellow-brown to reddish brown hypothecium. Photo from southern ON.

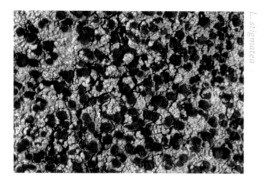

L. stigmatea

Lepraria
Dust lichens

Crustose lichens with a leprose thallus, usually consisting only of patches of soredia, sometimes clustered into granular aggregations. Sterile, without apothecia or perithecia. Thallus margins can be vague or distinct. One species, *L. xerophila* (not pictured), forms marginal lobes and can appear somewhat squamulose; it is whitish to pale yellowish gray and occurs in coastal areas, Marin Co. to Mexico.

Lepraria finkii
Fluffy dust lichen

Thallus leprose, yellowish green to green or whitish, somewhat thick and cottony, with fine to coarse, fluffy soredia; without true lobes but sometimes

L. finkii

with marginal sublobes. Spot tests: K+ yellow or K–, KC–, C–, P+ orange. On shady rocks, moss, and tree bases statewide, mostly coastal. Resembles *L. pacifica*, which often grows on redwood trunks. Compare also with *L. neglecta*. Photo from southern ON.

Lepraria neglecta
Zoned dust lichen

Thallus bluish gray with coarse, granular, usually pruinose soredia, often forming rosettes, sometimes in concentric rings when on rock. Spot tests variable, as 5 chemotypes have been distinguished. Mostly on soil and mossy rock in coastal areas and mountains, statewide. Probably the most common species of *Lepraria* in central and southern CA. Tends to be replaced by *L. finkii* from Marin Co. northward. Photo from San Luis Obispo Co.

L. neglecta

Lepraria pacifica
Pacific dust lichen

Thallus leprose, green, or white to pale blue, forming rosettes or irregular patches; soredia coarse and often pruinose. Spot tests: K+ yellow, C+ red, KC+ red, P+ yellow, UV+ dull whitish. On rock, soil, and bark, statewide. One of the few lichens commonly found on the lower trunks of redwoods. Photo of the lichen on a redwood trunk, Marin Co.

Lobothallia
Puffed sunken disk lichens

Foliose thalli that become crustose and areolate in the center but retain a lower cortex. Adnate, with thick, flat to convex, marginal lobes 0.5–1 mm wide. Apothecia are dark brown to black, on the surface or sunken. Spores broadly ellipsoid, colorless, 1-celled, 8 per ascus. Spot tests: cortex and usually medulla K+ red, KC–, C–, P+ yellow. Can resemble some species of *Aspicilia*, except that the apothecia are lecanorine and *Aspicilia* species are not

generally as lobate; it can also be mistaken for some species of *Lecanora* or *Protoparmelia*. On rock in mountains and deserts.

Lobothallia alphoplaca
Variable sunken disk lichen

Thallus thick, variable in color from creamy gray to brown, with radiating, very convex lobes. Apothecia usually numerous, dark brown, crowded near the center. On rock, central to southern CA mountains and

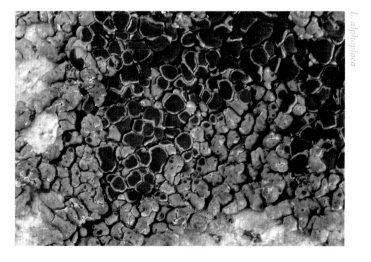

Mojave Desert. Compare with *L. praeradiosa*. Photo from the Mojave Desert near Twentynine Palms.

Lobothallia praeradiosa

Thallus yellowish gray to copper brown, can be areolate in the middle but typically lobate at the edges with flat lobes, tightly attached to the rock substrate. Apothecia brown to black. Similar habitat and range to *L. alphoplaca*, but yellower, flatter, and more tightly attached. Photo from near Moab, UT.

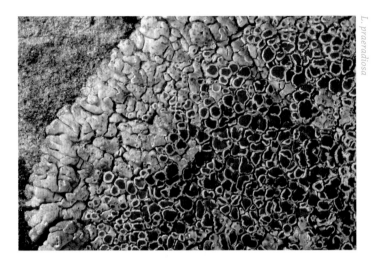

L. praeradiosa

Megaspora verrucosa

False sunken disk lichen

Crustose lichen with a whitish, fairly thick, cracked and lumpy thallus. Apothecia black, sunken, sometimes with lecanorine margins. Spores broadly ellipsoid, colorless, 1-celled, thick-walled, 8 per ascus. Spot tests negative, but the greenish epihymenium intensifies with K. Occasional on bark, plant

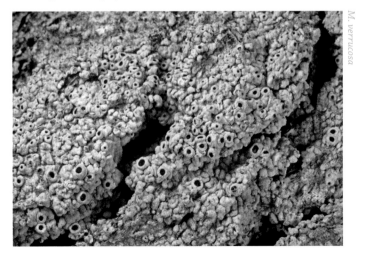

M. verrucosa

debris, or soil, typically in mountain areas, statewide. Somewhat resembles *Diploschistes*, or *Aspicilia cyanescens*, the only species of *Aspicilia* that grows on bark and wood, but that differs in having thicker spore walls and branched paraphyses that are expanded into knobs or beadlike cells at the tip, as in other species of *Aspicilia*. Photo from Mendocino National Forest.

Miriquidica scotopholis
Brown dragon scale
Thallus crustose, areolate to squamulose, with small, dark brown, shiny, overlapping areoles like tiny scales. Apothecia common, small, scattered, 0.5–1 mm wide, very dark brown to black, lecanorine when young, becoming lecideine with maturity. Spores colorless, oblong, simple, 8 per ascus. Spot tests negative. Somewhat resembles *Lecidea atrobrunnea*, but differs in microscopic structure and chemistry. Also compare with *Rhizocarpon bolanderi*. Occasional on rock, most often in coastal mountains, Mexico to Marin Co., possibly farther north. Photo from San Luis Obispo Co.

M. scotopholis

Mycoblastus sanguinarius
Bloody-heart lichen
Thallus thick, crustose, white, or pale green to yellowish, continuous to verruculose, often growing over moss and other lichens. Apothecia black, irregular, up to 2.5 mm wide, without margins, with a bright red zone below the brown hypothecium. The color will distinguish this lichen from similar-looking ones, such as species of *Lecidella*. In addition, the large, 1-celled, colorless spores of *Mycoblastus* have very thick walls. Spot tests: thallus K+ yellow, KC–, C–, P–; red pigment K+ bright red-orange. Occasional on bark and wood, northwestern CA. Photo from a herbarium specimen collected in OR.

Normandina pulchella

Elf-ear lichen

Thallus small, blue-green to grayish, squamulose, forming irregular patches with tiny rounded lobes 1–2 mm wide that usually develop raised, sorediate rims; soredia sometimes laminal as well. Lower surface white, without a cortex. Fruiting bodies are perithecia, but very rare. Spot tests negative. Uncommon, on moss and other lichens, especially those containing cyanobacteria like *Fuscopannaria* or *Peltigera*, occasionally on bark or rock, in humid locations coast and inland, statewide. Photo of a damp specimen from the Nantahela National Forest, NC.

Ochrolechia
Saucer lichens, Cudbear

Crustose lichens with thalli that are white or pale gray. Round apothecia have a thick, lecanorine margin around an orange, tan, or pinkish disk. Spores colorless, 1-celled, thin-walled, 4–8 per ascus. Can resemble some species of *Lecanora*, but those have smaller spores, mostly unbranched paraphyses, and seldom react C+. When making chemical tests on *Ochrolechia* species, it is usually important to distinguish between reactions in the cortex and those in the medulla. Almost all species grow on bark or wood. The name "Cudbear" comes from Scotland, where a species of *Ochrolechia* was formerly used to dye wool. A few additional species, not pictured (none common), are found in CA; one of them, *O. pseudopallescens,* is unusual in growing on rock.

Ochrolechia africana
Frosty saucer lichen

Thallus usually bumpy, pale yellowish gray, often showing a shiny white prothallus. Apothecia typically under 1.5 mm across, yellowish pink to pale orange, pruinose, with very thick margins. Spot tests: thallus usually UV+ yellow; cortex of thallus and apothecial margin C–, KC–, but apothecial disk and medulla of apothecial margin C+ red, KC+ red. Occasional, on bark, sometimes on wood, mostly in coastal southern CA, but also reported from the northern Coast Range. A similar but rarer species on bark along the central and southern coast, *O. mexicana* (not pictured), reacts KC+ red, C+ pink to red in both cortex and medulla. Photo from a herbarium specimen collected in Baja CA.

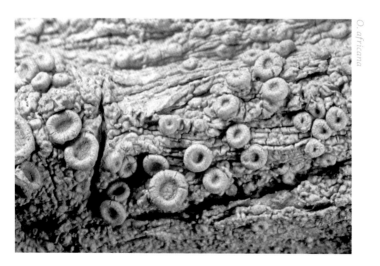

O. africana

Ochrolechia androgyna
Powdery saucer lichen

Pale gray, rugose to verrucose-areolate thallus, often thick, with granular, yellowish soredia in round clumps that become more dispersed and irregular in older parts of the thallus; apothecia rare. Spot tests: thallus cortex K– or pale yellow, KC+ red, C+ pink or red, P–; medulla K–, C+ pink or C–, KC–, P–; soralia C+ red, KC+ red; all tissues UV– or white. *Ochrolechia arborea* is also sorediate, but the thallus is thinner, and the soralia react UV+ bright yellow. On bark, wood, rock, and moss in the central and southern coastal ranges and Sierra Nevada. Photo from a herbarium specimen collected in NB.

O. androgyna

Ochrolechia arborea
Tree saucer lichen
Thallus thin, especially at the margins, and white, often forming circular patches, with pale green, usually round soralia containing granular soredia. Spot tests: thallus cortex and soralia K–, KC+ red, C+ red, P–, thallus UV– or

O. arborea

bright yellow in part, soralia UV+ bright yellow. Resembles *O. androgyna*, above. Occasional on bark and wood of both deciduous and conifer trees, in coastal habitats, San Francisco to Mexico. Photo from southern ON.

Ochrolechia juvenalis
Juvenile saucer lichen
Thallus thin, pale gray, continuous at the edges but verruculose in the center. Apothecial disks round, varying in size, starting out small like pores but later broadening, yellowish pink, heavily pruinose, with thick margins. Spot tests: cortex K–, KC–, C+ yellow or C–, UV–; disk K+ yellow, KC+ red, C+ pink (under the pruina); medulla no reactions. Occasional on conifer bark, in the Coast Range and interior northern mountains, Monterey Co. to the OR border. Photo from a herbarium specimen collected in BC.

O. juvenalis

Ochrolechia laevigata
Smooth saucer lichen
Thallus very thin, continuous, pale yellowish or pinkish gray to white. Apothecia up to 4.5 mm wide, light orange, with smooth margins, epruinose. Algae sparse in apothecial margins. Spot tests: mostly negative, but apothecial disk and marginal cortex, and part of the medulla, C+ red, UV–. Occasional on bark of deciduous trees, especially red alder, sometimes on maple, rarely on

conifers, in northern coastal forests. Compare with *O. oregonensis* and *O. sub-pallescens*. Photo from Del Norte Co.

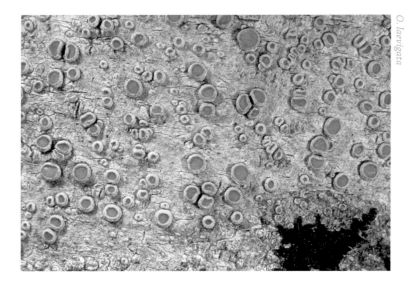

O. laevigata

Ochrolechia oregonensis
Double-rim saucer lichen

Thallus thick, verrucose, yellowish white. Apothecia large, up to 4 mm wide, often developing double margins around an orange disk; the inner ring is smoother and more disk-colored, the outer one is often broken and typically white like the thallus. Spot tests: apothecial disk and margin, and thallus cortex C+ red, UV–, medulla C–. On conifer bark, especially Douglas-fir, mostly in the Coast Range, San Luis Obispo Co. northward, but occasionally in northern interior ranges and southern mountains. Photo from northern CA or southern OR.

O. oregonensis

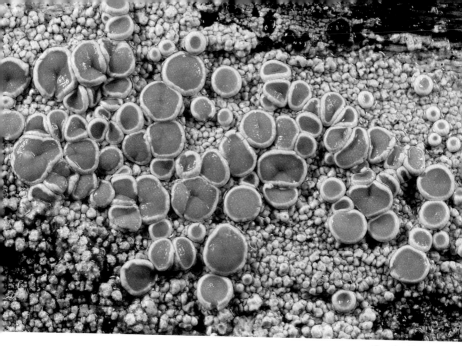

Ochrolechia subpallescens

Thallus thin to moderately thick, verruculose, yellowish gray to white, epruinose. Apothecia orange with white margins. Algal layer continuous under the hypothecium. Spot tests: apothecial disk K–, KC–, C+ red, P–; thallus cortex and apothecial margin K–, KC+ red, C+ red, P–; medulla no reactions; all tissues UV–. Mostly on deciduous trees, occasionally on conifers or on wood, fairly common in coastal forests statewide, occasional in interior mountains. *O. subpallescens* has a thicker and bumpier thallus than *O. laevigata,* and lacks the double margin of *O. oregonensis.* The photo, from a fence in coastal Marin Co., shows a very wet specimen, but this species doesn't change color much when moist.

Ochrolechia upsaliensis

Tundra saucer lichen

Thallus somewhat thick and granular, yellowish. Apothecia abundant, pale yellow, coarsely pruinose, 0.6–2 mm wide, occasionally darkening to reddish purple, with prominent white margins. Spot tests: thallus C–, KC–; apothecial disks, margin cortex, and medulla C– or yellow, KC– or yellow; apothecial disks UV– or yellow. Occasional, on moss, other vegetation, and soil, in the Coast Range, San Benito Co. northward, and in northern interior mountains. Frequently found growing over *Selaginella.* A mostly arctic-alpine species. Photo from Olympic National Park, WA.

Opegrapha

Scribble lichens

Crustose lichens with thalli that are thick or thin, sometimes inside the substrate and not visible, with elongated black fruiting bodies (lirellae), similar in appearance to *Graphis*. Paraphyses are branched and anastomosing, unlike other script lichens, and cells of the spores are square, not lens-shaped as in *Graphis*. Spot tests negative for most species. On bark and rock. A number of *Opegrapha* species, not described below, occur in CA. Of these, *O. anomea* is unusual in being parasitic on other lichens, especially *Ochrolechia* and *Pertusaria* along the coast from Monterey Co. to Mexico. The species that used to be named *Opegrapha atra* is now called *Arthonia atra*.

Opegrapha herbarum

Plant scribble lichen

Thallus thin or immersed, smooth pale gray to brownish or olive, with scattered ascomata, simple or branched, black, with elongated slits. Pycnidia

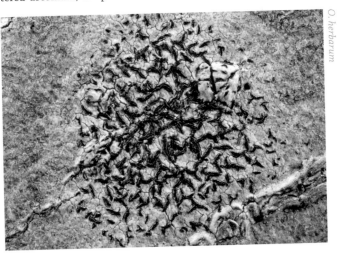

rare. Spores ellipsoid to fusiform, colorless but becoming brownish with age, 4-celled, typically 18–28 × 5–8 μm. On smooth bark or dead leaves, sometimes on rock, in coastal habitats, San Mateo Co. to Mexico. One of the more common species of *Opegrapha*. Photo from a fig tree in coastal San Luis Obispo Co.

Opegrapha vulgata

Thallus thin, smooth, pale gray to brown, with black, somewhat immersed ascomata like tiny slits but short, and irregular in shape. Often with numerous pycnidia mixed with the lirellae. Spores narrowly fusiform and a bit tapered at one end, straight or slightly curved, colorless, usually 5–8-celled, typically 19–30 × 3–4 μm. A similar species, *O. niveoatra* (not pictured), has shorter conidia. Occasional on bark, especially cypress, coastal southern CA. Photo from coastal MA.

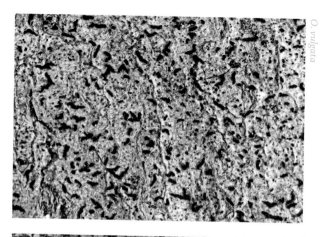

O. vulgata

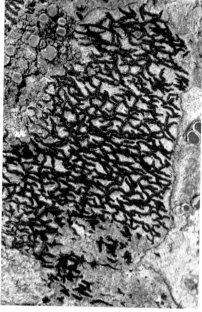

This is an unidentified species of *Opegrapha* from the coast of Humboldt Co., growing on alder bark. To determine a species of *Opegrapha*, one needs to examine the details of its spore structure.

Ophioparma rubricosa
Pacific bloodspot

Thallus crustose, thin to thick, yellowish and granular. Apothecia scarlet to rusty red, 1–3 mm wide, epruinose. Spores fusiform, colorless, 4–7-celled, 8 per ascus. Spot tests: K–, KC+ yellow, C–, P–. Rare, on wood, especially manzanita, or on conifer bark, northern Coast Range. The color of the apothecia is distinctive. Photo from southwestern OR.

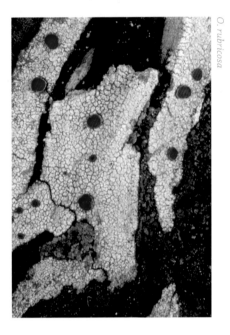

Peltula
Rock-olive lichens

Small, dark brown to olive or dark gray squamulose lichens with flat to erect, somewhat rounded lobes that attach with a central holdfast or a tiny cluster of rhizines. Apothecia are frequent in most species, sunken into the lobes and usually concave. Photobiont is the cyanobacterium *Anacystis*. Spores tiny, colorless, 1-celled, up to 100 per ascus. Spot tests usually negative. On soil and rock. *Peltula* species somewhat resemble those of *Psora* or *Placidium* but those genera have a green alga as the photobiont (and Placidium species have perithecia); or *Heppia,* which has a different cyanobacterium (*Scytonema*) and 8 spores per ascus.

Peltula bolanderi
Lobes dark gray to dark brown or olive, up to 2 mm wide, with wavy margins and black, farinose, marginal soralia. Apothecia rare. Inconspicuous, on rock in dry habitats both along the coast and inland, mostly southern CA, but occasional as far north as Mendocino Co. Photo from Marin Co. The tan lobes on the left side of the photo are those of *Peltula euploca*.

Peltula euploca
Powdery rock-olive

Thallus small, made up of irregularly rounded, umbilicate, brown, often somewhat scattered squamules up to 12 mm wide with a pale gray fringe of fine soredia. Apothecia rare. Distinctive and fairly common, on rock mostly in dry habitats, but can be coastal as well, statewide. Photo from Mariposa Co.

Peltula obscurans
Common rock-olive

Squamules dark olive to dark gray, usually somewhat scattered but sometimes forming small rosettes. Apothecia frequent, black, each one centered in a lobe. When wet, the apothecia turn reddish brown and the lobe surface becomes olive-green. On rock and soil in desert environments, mostly in southern CA, including the Mojave Desert, but as far north as inland Monterey Co. The photo shows variety *hassei,* the most common form in CA; var. *obscurans* and var. *deserticola* also occur in southern CA but more rarely. Photo from Santa Barbara Co.

P. obscurans

Peltula patellata
Stuffed rock-olive

Squamlues clustered, round to somewhat lobed, dark olive to brown, 1–4 mm wide. Apothecia bright orange to reddish brown, centered in the lobes, sometimes small or can almost cover the lobe, sometimes also with thin margins. Epihymenium reacts K+ red-violet. Fairly common on soil in desert areas, southern CA. Compare with *P. richardsii.* Photo from inland San Luis Obispo Co.

P. patellata

Peltula richardsii
Giant rock-olive

Thallus forms patches of round, rather large, olive or brown squamules with downturned margins, 2–10 mm wide. Apothecia dark red, centered in the lobes, typically without margins, usually conspicuous but can be immersed and barely visible. Epihymenium reacts K+ red-violet. On calcareous soil in deserts, southern CA. The epihymenium reaction with K is the same as in *P. patellata,* but that species has squamules with upturned margins, a pale brown lower surface, and conspicuous and persistent apothecial margins. Also resembles the rarer *Heppia conchiloba,* but that species has 8 spores per ascus, and its spot tests are all negative. Photo from southern AZ.

P. richardsii

Peltula zahlbruckneri
Squamules olive to dark gray-brown with somewhat swollen lobes that can overlap and become almost fruticose; usually in irregular patches. Apothecia

P. zahlbruckneri

immersed, reddish brown to almost black, lacking margins, 1–7 per squamule. On acidic rock, especially granite, in arid habitats, scattered statewide, but more common in southern CA. Photo from Mariposa Co.

Also of Note
Another CA species is *Peltula omphaliza* (not pictured), with dull, olive to brown or black squamules, often scattered; apothecia small, immersed, yellowish, 5–20 per squamule; spot tests negative; on rock, mostly in southern mountains.

Pertusaria
Wart lichens
A crustose genus with many diverse species, typically with fruiting bodies that are modified apothecia immersed in warts on the thallus that either open in small pores like ostioles or are fully open and look like lecanorine apothecia. Some species are coarsely sorediate. Spores large, colorless or occasionally light brown, 1-celled, usually thick-walled and layered, 1–8 per ascus. On all substrates. A few other species, not described below, occur in CA. Perhaps the most common is *P. albescens*, with a thick, pale bluish gray thallus; without apothecia but with large, disklike soralia; spot tests: K–, C–, KC–, P–, UV–; on bark, central and southern coastal mountains.

Pertusaria amara
Bitter wart lichen
Thallus thin, dark or pale gray to greenish gray, with milk white soralia that contain coarse, granular soredia. Thallus margin often scattered with small, white dots (pseudocyphellae). Apothecia absent. Spot tests: K– or K+ yellow to red, KC+ purple, P– or yellow to red. The lichen is fairly easy to identify by the extremely bitter taste of the soredia (taste only what can fit under a fingernail), which contain picrolichenic acid. On bark, occasionally on rock, quite common statewide near the coast, occasional in southern mountains. Photo from the Channel Is.

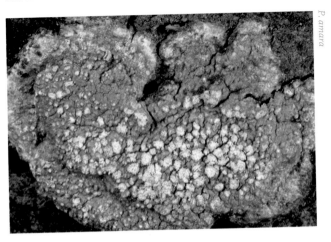

P. amara

Pertusaria californica
California wart lichen

Thallus whitish to dark gray, areolate, with large, thick verrucae, each containing 1–10 small, dark brown to black, immersed apothecia. Spores 1–2 per ascus. Spot tests: K+ yellow to red, KC–, C–, P+ yellow to orange. On coastal rocks, statewide, sometimes forming large patches. Superficially resembles the fertile form of *Dirina catalinariae*. Photo from Marin Co.

P. californica

Pertusaria flavicunda
Sulphur wart lichen

Thallus areolate, with fairly thick verrucae, pale yellowish. Apothecia numerous, lecanorine, black but appearing lighter due to a coating of yellowish pruina. Occasionally with patches of soredia as well. Spores 2 (occasionally 3) per ascus. Spot tests: cortex K– or yellow, KC+ orange, C+ pale orange, P– or

P. flavicunda

pale orange; medulla K– or faint yellow, KC– or spotty orange, C–, P– or pale orange. The yellowish color and pruinose, lecanorine apothecia make the species fairly easy to recognize. On rock, commonly sandstone, along the coast and in nearby mountains, Santa Barbara Co. south. Photo from the Channel Is.

Pertusaria ophthalmiza
Ragged wart lichen

Thallus gray, usually smooth and thin, with fruiting warts that have wide, black, but coarsely white-pruinose disks and fragmented margins that make the disks look sorediate; some authorities characterize the disks as sorediate. Spores 1 per ascus. Spot tests negative. Uncommon, on bark, in mostly coastal habitats, Los Angeles Co. to OR. Resembles *P. subambigens*, which is 8-spored and PD+ red, and *P. amara*, which has KC+ purple granular soredia on the warts. Photo from Mt. Baker–Snoqualmie National Forest, WA.

P. ophthalmiza

Pertusaria rubefacta
Bumpy wart lichen

Thallus pale greenish yellow to gray, with moderately thick, rounded verrucae containing 1–5 sunken apothecia that are dark brown to almost black but sometimes covered with gray pruina. Ascomata at first are ostioles, then they dilate and fuse to form broader disks resembling lecanorine apothecia. Spores 8 per ascus. Spot tests: medulla K+ yellow to red, P+ yellow to orange; cortex C+ orange, KC–, P+ yellow to orange, UV+ orange-red; epihymenium K+ violet. On the bark of broadleaved trees, especially oaks. Occasional in coastal areas from Santa Cruz Co. to Mexico. A similar, but rarer, species is *P. hymenea* (not pictured), which reacts K– or K+ yellow, C– or C+ yellow, KC+ orange, P–, UV+ orange-red; on bark near the coast, central and southern CA. Photo from northern ME.

Pertusaria subambigens
Frosted wart lichen

Thallus gray to greenish, delimited at the margin. Fruiting warts each contain an apothecium whose color varies from yellow to pink or green but appears whitish due to a heavy coating of pruina. Apothecial margins rough, forming concentric rings. Spot tests: medulla and apothecial disk K–, KC–, C–, P+ red. The ragged margins of the apothecia, broader disks, and the P+ disk reaction distinguish it from *P. ophthalmiza*. Mainly on conifers, central to northern Coast Range. Photo from Willamette National Forest, OR.

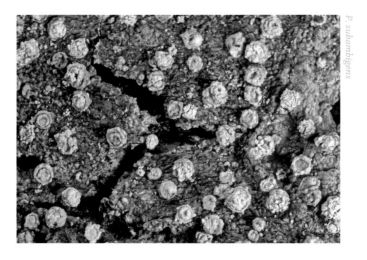

Pertusaria velata
Rimmed wart lichen

Thallus pale gray to yellowish, thin at the margin but thick and verrucose in the center. Apothecia lecanorine, pale yellow to pink, disks covered with heavy, coarse, white pruina. Spot tests: cortex all reactions negative except sometimes UV+ yellow; medulla and apothecial disk K–, KC+ red, C+ deep red, P–, UV–. Occasional on bark, especially oaks, sometimes on mossy rock, in coastal environments, Marin Co. to Mexican border. Photo from Gold Head Branch State Park, FL.

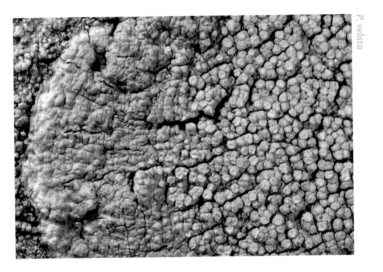

P. velata

Pertusaria xanthodes
Volcano wart lichen

Thallus thin to somewhat thick, pale yellowish gray to greenish, rough and cracked or continuous. Fruiting warts with sloping sides and sunken apothecia appearing as small, pale ostioles, but extremely variable. Spores have two

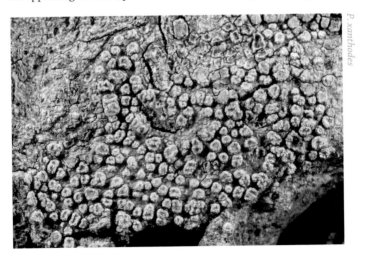

P. xanthodes

thick walls with the inner layer "rough," 2 per ascus. Spot tests: cortex KC+ orange, UV+ orange, medulla K+ yellow, P+ yellow to orange. Fairly frequent on deciduous trees in coastal zones, Marin Co. to Mexico. A very similar species, *P. pustulata* (not pictured), has a gray thallus and black ostioles; fairly common on bark in coastal forests, Sonoma Co. to Mexico. Also resembles *P. leioplaca* (not pictured), with a grayish to light green thallus, mostly epruinose, with fertile verrucaea and 4 spores per ascus; fairly common on bark in central and southern coastal habitats. Yet another species that can resemble *P. xanthodes* is *P. lecanina* (not pictured), with a yellowish gray, verrucose thallus, pruinose apothecia, and 2-spored asci; fairly common on bark along the central and southern coast. Photo from San Luis Obispo Co.

Phlyctis argena
Whitewash lichen

Thallus pale greenish or yellowish to white, verruculose to granular, rough or smooth, consisting of tiny warts that break down into patches of coarse granular soredia, sometimes encompassing the entire thallus;. Apothecia rare. Spores huge, muriform, 1 per ascus. Spot tests: K+ yellow becoming blood-red, KC–, C–, P+ deep yellow. When just a mass of white soredia, it could be mistaken for a species of *Lepraria*. Occasional, on bark, sometimes on rock, in central Coast Range. Photo from Butte Co.

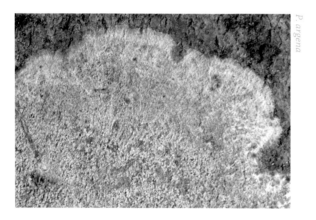
P. argena

Placidium
Stipplescale lichens, Earthscale

Squamulose lichens that can become almost foliose or sometimes look like a lumpy crust, gray to dark brown but often becoming greener when wet. Fruiting bodies are perithecia, buried in the lobes and appearing only as black dots on the surface. Lower surfaces lack rhizines, unlike the genus *Clavascidium*. Spores ellipsoid to almost globose, colorless, 1-celled. Spot tests negative. Most species grow on soil, occasionally on rock; only the rare *P. arboreum* grows on bark. Found mostly in deserts and high mountains.

Placidium acarosporoides

Mojave stipplescale

Thallus squamulose, reddish to dark brown or black, usually shiny, with convex, discrete or continuous lobes dotted with black perithecia, one to several per squamule. On granitic rock and sandstone in southern deserts. Photo from the Mojave Desert.

P. acarosporoides

Placidium arboreum

Tree stipplescale

Thallus of rounded tan lobes, 2–5 mm wide, that become bright green when wet, spotted with dark brown perithecia. Lower surface pale, with attachment

P. arboreum

hyphae in tufts. On bark, especially oaks, in open woodlands; rarely on mossy rock. Rare in CA, but has been reported from a few Coast Range locales. Photo from Big Bend National Park, TX.

Placidium lachneum
Mountain stipplescale

Thallus very similar to *Clavascidium lacinulatum,* but that species has rhizines; *P. lachneum* has a black lower cortex with vertically arranged columns of cells. Uncommon, on soil, moss, and plant debris, found farther north than *C. lacinulatum,* usually in mountain ranges, Coast Range and inland. Photo from the alpine Sierra Nevada, Inyo Co.

P. lachneum

Placidium squamulosum
Squamule stipplescale

Thallus very similar to *Clavascidium lacinulatum,* but lower surface has fewer hyphae and lacks true rhizines. Widespread and fairly common, on soil,

P. squamulosum

especially if calcareous, sometimes on moss or plant debris, in arid habitats, especially southern mountains, but also coastal. Also resembles a rarer species from the southern coast and deserts, *P. pilosellum* (not pictured), which has marginal pycnidia. Photo from the Channel Is.

Placopsis lambii
Pink bull's-eye lichen

Thallus crustose, forming a rounded rosette with lobed margins; white to pinkish, or light yellowish brown, becoming greener when wet. Usually with a pinkish tan cephalodium in the middle of the thallus and smaller cephalodia scattered on the thallus surface; also producing rounded clumps of green or brownish gray soredia, and often tan apothecia that become bright pink when wet. Spores ellipsoid, colorless, 1-celled, 8 per ascus. Spot tests: K–, KC+ red, C+ pink, P–. Rare, on rock, northwest coast. A similar, but even rarer species, *P. gelida* (not pictured), is usually darker and browner, with deeper cephalodial lobes and irregular, not hemispherical, soralia. The photo, from western OR, shows a damp, but not soaking wet, thallus.

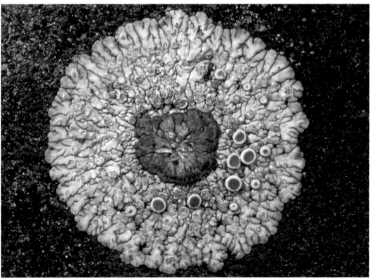

Placynthiella uliginosa
Tar-spot lichen

Thallus thin, crustose, granular to granular-isidiate, dark brown to greenish. Apothecia dark red-brown to almost black, lecideine. Spores ellipsoid, colorless, 1-celled, thin-walled, 8 per ascus. Spot tests negative. Widespread but uncommon, on soil or rotting wood, never on bark, sometimes on sandy soil, In the Coast Range, southern mountains, and Sierra Nevada. Quite similar, and sharing similar habitats, is the less common *P. oligotropha* (not pictured), which has a thallus with much larger granules, up to 0.3 mm wide. Photo from NB.

Placynthium nigrum
Ink lichen

Thallus crustose, dark olive to black with a distinctly bluish prothallus, continuous or areolate, composed of masses of tiny squamules often covered with isidia. Apothecia flat, or slightly convex, lecideine. Spores narrowly ellipsoid, 2–4-celled, 8 per ascus. Spot tests negative. Occasional, on calcareous rock, often on concrete, statewide. A rarer species, *P. asperellum* (not pictured), lacks the prothallus of *P. nigrum* and has long, narrow, marginal lobes and slender isidia. Photo from the Chiracahua Mountains, southern AZ.

Pleopsidium flavum
Common gold cobblestone lichen

Thallus brilliant yellow or yellow-green, crustose, often with radiating, convex, and folded lobes at the margin, becoming areolate in the central portions, with a rather rough surface. Apothecia frequent, dark yellow to brownish, less than 1 mm wide, with thalline margins. Spot tests negative except UV+ orange. Fairly common, on rock in dry habitats, central and southern CA. Resembles *Acarospora socialis,* but *P. flavum* is flatter, with an effigurate margin, and is more montane. Photo from inland Monterey Co.

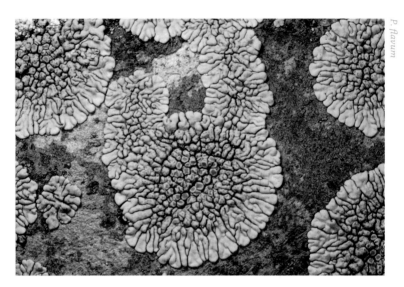

P. flavum

Polysporina simplex
Common coal-dust lichen

No visible thallus, only black, lecideine apothecia on rock, often in groups that follow the cracks. The apothecia, 0.3–1.0 mm wide, have a rough, brittle, carbonized margin and epihymenium, and sterile black bumps on the disks. Hypothecium colorless. Paraphyses slightly branched and anastomosing, not wider at the tips. Spores tiny, elliptical to cylindrical, hundreds per ascus. Spot tests: hymenium IKI+ reddish orange, K/I+ blue; ascus walls K/I–. Fairly common in central and southern CA, especially near the coast, but also in the Sierra Nevada. Compare with *Sarcogyne clavus* and endolithic species of *Lecidea,* such as *L. laboriosa.* A similar species in the southern deserts and Sierra Nevada, *P. subfuscescens* (not pictured), has larger, more convex apothecia and typically grows on other lichens. Photo from near Escalante, UT.

Porpidia crustulata
Concentric boulder lichen

Thallus crustose, thin, pale gray to greenish, continuous or a bit cracked. Apothecia black, lecideine, often forming concentric rings. Spores colorless, 1-celled, 8 per ascus. Spot tests negative. Could easily be confused with a

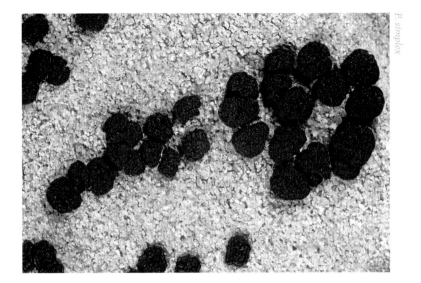

species of *Lecidea* or *Buellia,* but *Porpidia* species have a distinctive ascus structure forming a deeply blue-staining "plug" or tube in the otherwise lightly stained ascus tip when treated with iodine. Occasional on rock, northern Coast Range. Photo from Ouachita National Forest, AR.

Protoparmelia badia
Chocolate rim lichen

Thallus crustose, areolate to verrucose, brown to grayish brown or olive, sometimes poorly developed. Apothecia numerous, lecanorine, with dark brown, rather shiny disks that have thalline margins. Spores pointed, colorless, 1-celled, 8–16 × 3–8 μm, 8 per ascus. Spot tests: K– or K+ yellowish, KC+ red-violet but transient, C–, P–. Resembles a species of *Lecanora,* but

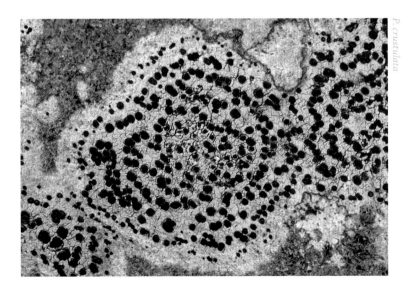

the coloration is distinctive. Occasional on rock, especially granite, in the Sierra Nevada and southern desert mountains. Another, rarer CA species is *P. ryaniana* (not pictured), with a brown, shiny thallus and dark brown to black apothecia; spores typically 6–10 × 2.5–3 μm; spot tests: K–, KC+ pink to red-violet, C–, P–, I–; initially a parasite on other lichens, becoming independent on rock, central and southern coast. A bark-dwelling species of the genus, *P. ochrococca* (not pictured), is composed of shiny red-brown granules, usually without apothecia, and is found on conifers in the mountains. Photo from the Rocky Mountains, CO.

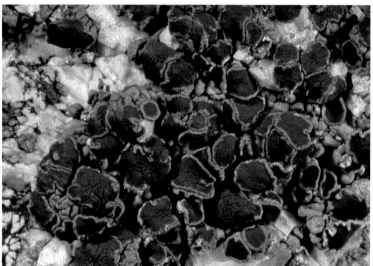

P. badia

Psora

Scale lichens

Small squamulose lichens with fairly large, thick lobes, 2–6 mm wide, most often brownish and turning green when wet, but some species pale gray or pink, and some with pruina. Lower surface pale, with a cortex, and tomentum that attaches the thallus to the substrate. Often fertile, with biatorine, convex or spherical, reddish brown to black apothecia. Spores ellipsoid, colorless, 1-celled, 8 per ascus. Usually on soil, or thin soil over rock, especially if calcareous, most often in somewhat arid habitats. Compare with *Placidium* species, which have perithecia, or *Heppia* and *Peltula,* which contain cyanobacteria.

Psora californica

Thallus with brown, shiny, epruinose squamules with downturned margins and mostly without fissures. Apothecia dark brown to black. Spot tests: upper cortex C+ red, KC+ reddish (can be difficult to see), medulla C– or C+ pink. In mountains, most often in the central and southern Coast Range and Sierra Nevada. Photo from a herbarium specimen collected in San Bernardino Co.

Psora cerebriformis
Brain scale

Thallus squamulose, pale yellow-brown to olive, but usually with a coating of white pruina, with convex or flat lobes 2–8 mm wide that often cluster to form a rounded lump; the lobes are fissured, giving it a brainlike appearance. Apothecia frequent on the margins, black. Spot tests: cortex and medulla K+ yellow or red, KC–, C–, P–. Occasional on soil in dry habitats, eastern side of the Sierra Nevada and in northeastern CA. A desert and Great Basin species. Photo from Grand Canyon National Park, AZ.

Psora crenata

Brick scale

Squamules separate or overlapping, up to 10 mm wide and usually concave but with downturned margins, reddish but almost always covered in a heavy white pruina. Apothecia abundant on the margins, black but sometimes appearing tan to somewhat red because of the coating of pruina. Spot tests: medulla K+ red, KC–, C–, P+ yellow to orange. Occasional on soil in deserts, desert mountains, and juniper-pinyon pine woodlands, southern CA. Photo from Pedernales Falls State Park, TX.

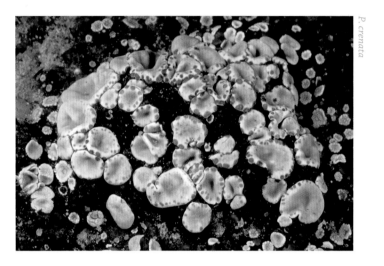

P. crenata

Psora decipiens

Blushing scale

Squamules pink-red to orange-red, usually with white pruina around the margins and sometimes covering the entire lobe, 1–6 mm wide, flat or somewhat upturned and often with frayed edges. Apothecia black, hemispherical,

P. decipiens

on the margins of some squamules. Spot tests negative. On soil, especially when calcareous, rarely on rock, in deserts and dry mountain habitats, central and southern CA. Similar to *P. crenata*, but with flat to convex squamules and upturned lobe margins. Photo from UT.

Psora globifera
Blackberry scale

Squamules reddish brown, sometimes yellowish or greenish, 2–5 mm wide, often shiny and usually epruinose, crowded and overlapping, with upturned margins, often with fissures in older squamules. Apothecia medium brown to black, slightly convex to hemispherical, borne singly or in clusters on the surface but not on the margins of the squamules. Spot tests negative. Occasional, usually on rock, sometimes on soil, in drier locations, statewide, central to southern mountains. A similar, fairly common species is *P. luridella* (not pictured), with smaller, paler, more adnate squamules, and darker, epruinose apothecia; found in southern coastal mountains. Photo from Butte Co.

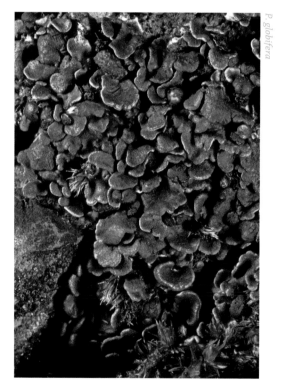

P. globifera

Psora nipponica
Butterfly scale

Squamules olive to greenish brown when dry, green when wet. Lobes broad, 2–10 mm wide, but fairly thin, rounded and overlapping, often fissured and curling inward, exposing the pale lower surface, with tan to white margins. Apothecia almost always present, dark brown to black, rounded, 1.5–2 mm wide, usually nestled among the squamules like tiny bunches of grapes. Spot tests: cortex and medulla K–, KC+ red, C+ pink, P–. On soil or rock, usually with moss, in the Sierra Nevada and Coast Range, statewide, and often in more humid, coastal locations than most species. One of the more common species of *Psora* in CA. Photo of a wet specimen from Mt. Tamalpais, Marin Co.

P. nipponica

Psora pacifica

Pacific scale

Squamules up to 3 mm wide, reddish brown when dry, turning green when wet, somewhat dispersed, irregular, loosely attached and ascending. The margins have tiny rounded lobules, and more erect lobes are somewhat pruinose.

P. pacifica

Apothecia rounded, reddish, without margins, present in the centers of many of the squamules. Spot tests: medulla K−, KC+ red, C+ pink, P−. Occasional on soil, central and southern coast and coastal mountains; rare in the Sierra Nevada. Photo from the Channel Is.

Psora tuckermanii
Brown-eyed scale

Squamules smooth, brown, usually partly pruinose, especially on the margins, scattered or crowded together and often overlapping. Apothecia common, laminal, reddish brown (less often dark brown to black), convex to hemispherical, without margins. Spot tests negative. On soil and rock, particularly sandstone, in both inland and coastal locations statewide. Somewhat similar in appearance is *P. russellii* (not pictured), which can be pruinose or not and has brown apothecia; spot tests: upper cortex and medulla K+ red, KC−, C−, P+ orange; on soil in southern CA deserts. Photo from UT.

Pyrenula occidentalis
Western pox lichen

A crustose lichen with the thallus mostly immersed inside the bark substrate, causing the bark to appear brownish yellow-green. Fruiting bodies are large, rounded, black perithecia, semi-immersed in the thallus and bark. Spores lens-shaped, brown, broadly fusiform, 4-celled. Spot tests negative, except the thin coating of orange pigment on the tips of some perithecia reacts K+ red-purple. Occasional on bark, especially such broadleaved trees as alder, along the northern CA coast. *Pyrenula* is a common genus with many species in the southeastern US; this is the only West Coast species. Photo from Humboldt Co.

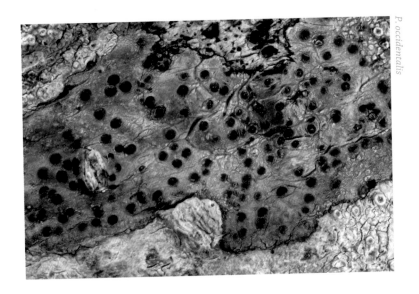

Pyrrhospora quernea

Sulphured crimson dot lichen

Thallus leprose, yellowish to pale greenish yellow, consisting entirely of granular soredia, spotted with black, or dark red, apothecia that vary in size and shape but are usually convex, with inconspicuous or absent margins. Spores broadly ellipsoid to ellipsoid, simple, colorless, typically 8–12 × 6–7 μm. Spot tests: thallus K+ yellow (but hard to see against the yellow thallus tissue), KC+ orange-red, C+ orange-red, UV+ orange or red. On bark, especially of oaks, but occasionally on wood or sandstone, in coastal locations statewide. Some taxa that used to be in this genus are now classified as species of *Ramboldia*. Photo from the Channel Is.; the lichen with gray apothecia around *P. quernea* in the photo is probably *Lecanora caesiorubella*.

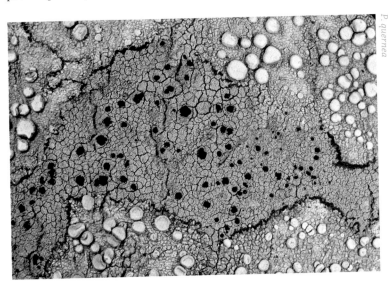

Ramboldia gowardiana

Northern crimson dot lichen

Thallus thin, pale grayish. Apothecia deep orange-red, appearing as irregular lumps, with thin or no margins. Without soredia. Spores ellipsoid, colorless, 1-celled. Spot tests: thallus K+ yellow or K–, KC–, C–, P+ red. Compare with *Caloplaca* species. Uncommon but distinctive, on bark, northern Coast Range. Photo from Southeast AK.

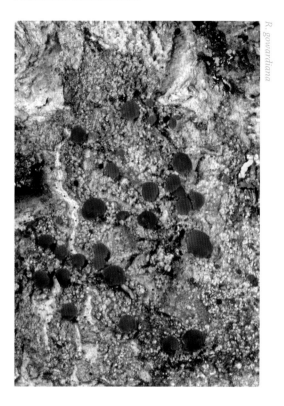

R. gowardiana

Rhizocarpon

Map lichens

Crustose lichens, most often with areolate thalli but can be continuous or rimose; usually with a black prothallus showing around the edge and sometimes between the areoles as well. The prothallus often divides adjacent thalli, giving the lichen a maplike look, especially in the bright yellow-green *geographicum* group. Apothecia black, lecideine. Spores usually muriform, colorless, brown, or greenish, 2- or more celled, 1–8 per ascus, with a gelatinous sheath. Widespread, on rock. Gray and brown species can resemble species of *Lecidea* and *Buellia,* and close examination of the spores is necessary to distinguish them; species with immersed apothecia may look like *Aspicilia,* but the exciple in that genus is poorly developed and colorless. CA has a number of other, less common, *Rhizocarpon* species, yellow-green and other colors, not described below.

Rhizocarpon bolanderi

Thallus with brown or red-brown peltate areoles slightly upturned at the margins, creating a gray rim around each areole, scattered or crowded on a thick black prothallus with black apothecia between the areoles. Spores dark, muriform, 2 per ascus. Spot tests: cortex negative; medulla usually K+ yellow. Common on rock in the Sierra Nevada and other interior ranges. It could be mistaken for a brown, squamulose species of *Lecidea* or for *Miriquidica scotopholis*. Photo from a herbarium specimen collected in the San Jacinto Mountains, Riverside Co.

R. bolanderi

Rhizocarpon disporum

Single-spored map lichen

Brown or pinkish to gray thallus usually with convex, often rounded areoles 0.2–0.6 mm wide, dispersed or continuous, usually with a black prothallus. Apothecia convex or flat, rounded or angular, with or without margins. One

R. disporum

spore per ascus, despite the scientific name. Spot tests: thallus usually K+ pale yellow but sometimes strong or absent, P– or reddish; medulla K– or red, KC–, C–, P– or yellow, IKI–. On rock, statewide, especially in inland mountains. Resembles *R. geminatum* (not pictured), a 2-spored species with a similar habitat and range. Compare also with *R. grande.* Photo from Coronado National Forest, AZ.

Rhizocarpon geographicum
Yellow map lichen

The most common of more than a dozen bright yellow to yellow-green species of *Rhizocarpon,* distinguished by spores and chemistry. Thallus areolate but quite variable, with continuous or dispersed areoles. Color variable as well, but always some shade of bright greenish yellow. Apothecia black, up to 1 mm wide, found between the areoles but can be hard to see against the black prothallus. Spot tests: cortex UV+ orange; medulla K–, KC–, C–, P+ yellow or P–, IKI+ blue. Common on rock statewide, especially in the Sierra Nevada and other inland mountain ranges, but sometimes coastal as well. Occasionally along streams and even underwater part of the time, but typically on fully exposed montane, siliceous rocks. Compare with *R. lecanorinum, R. macrosporum,* and *R. riparium.* Photo from interior AK.

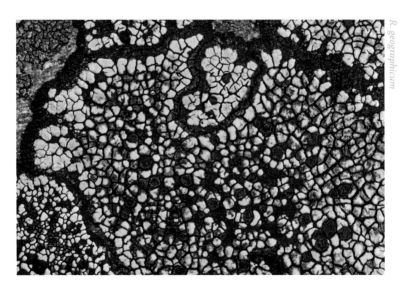

R. geographicum

Rhizocarpon grande
Gray map lichen

Thallus crustose, pale brownish gray to brown, with thick, somewhat rounded areoles 0.6–2 mm wide, usually with a black, sometimes faint prothallus. Apothecia black, found between the areoles. Spot tests: cortex C+ pale pink; medulla K+ yellow or K–, KC+ red, C+ red, P+ orange or P–. On acidic rock such as granite statewide in interior mountains. Similar to *R. disporum* but browner, with larger areoles and 8 spores per ascus. The photo, from the

Chiracahua Mountains, AZ, shows a specimen that is grayer than usual for the species.

R. grande

Rhizocarpon lecanorinum
Crescent map lichen

Thallus bright greenish yellow with convex, curved areoles that often partially surround black apothecia like the letter "C." A black prothallus is usually conspicuous around the edge and between the areoles. Spot tests: cortex UV+ orange; medulla K+ deep yellow, KC+ red or KC–, C+ pink or C–, P+ orange, IKI+ blue. The somewhat greener color than other, similar species and the crescent-shaped areoles make it distinctive. Fairly common, on exposed rock, statewide, mostly in interior mountain ranges. Compare with *R. riparium*. Photo from northern BC.

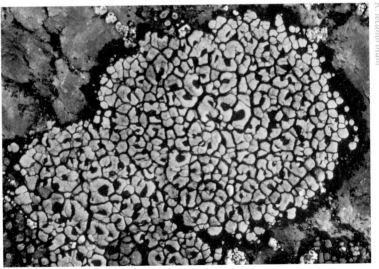

R. lecanorinum

Rhizocarpon macrosporum
Lemon map lichen

Thallus brilliant lemon-yellow to slightly greenish, with flat, irregular areoles that fit together like tiny tiles over a black prothallus. Apothecia black, single or clustered, sunken, found between the areoles. Spot tests: cortex UV+ orange, medulla usually K–, KC–, C–, P–, IKI+ blue-black. Resembles *R. geographicum,* except for the brighter color and larger spores (28–70 × 15–25 μm vs. 24–40 × 11–16 μm in *R. geographicum*). On rock in the Sierra Nevada, southern interior mountains, and northern Coast Range. Photo from White River National Forest, CO.

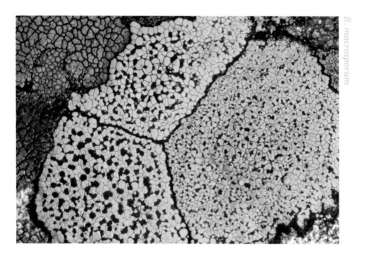

R. macrosporum

Rhizocarpon riparium
Streambank map lichen

Thallus yellow-green with irregular but somewhat rounded areoles, contiguous or dispersed, often with a conspicuous black prothallus. Apothecia black, sunken, up to 1 mm wide. Spot tests: medulla K– or K+ yellow or red, C– or

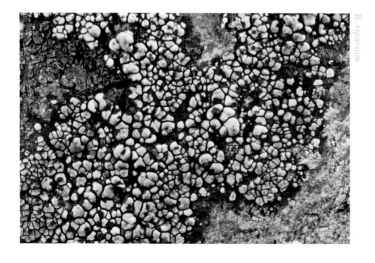

R. riparium

rarely C+ red, P– or P+ yellow. Similar in appearance to *R. geographicum,* but differs chemically and has both longer spores and a green hymenium. Statewide, especially on rock in mountain habitats. Also resembles *R. lecanorinum,* which has different chemistry and crescent-shaped areoles that partly surround the apothecia. Photo from Mt. Tamalpais, Marin Co.

Also of Note

One other CA species, *Rhizocarpon eupetraeum* (not pictured), has gray to brownish areoles, a distinct black prothallus, and black apothecia and pycnidia; spores dark brown, 22–42 × 10–18 µm, mostly 4 per ascus; spot tests: medulla K+ yellow to red, C–, KC–, P+ pale yellow; moderately common in interior mountain ranges.

Rinodina

Pepper-spore lichens
Small crustose lichens with gray to brownish or olive thalli that can be thick or thin. Apothecia are tiny, usually dark brown, lecanorine. *Rinodina* is a genus with many species; most are small and inconspicuous, rather like miniature versions of *Lecanora*. On all substrates, but most often on bark or wood. The various species are identified mainly by the structure of their apothecia and spores, especially the way the spore walls are thickened, and by chemistry. In particular, the spore types are important, and some of these are described in the glossary. Quite a few additional species, not described below, occur in CA.

Rinodina archaea

Primal pepper-spore lichen
Thallus thin or thick, rimose to areolate but often rather indistinct, gray to green or brownish. Apothecia dark brown to black with pale margins. Spores ellipsoid, brown, *Physconia*-type, 16–24 × 8–11 µm, 8 per ascus. Spot tests negative. Occasional on conifer bark, sometimes on wood, statewide, coast and inland mountains. Photo from Trinity Co.

Rinodina bolanderi

Bolander's pepper-spore lichen, Creamy pepper-spore lichen
Thallus variable, fairly thick, areolate to granular or verrucose, cream to yellowish gray or dark gray-brown. Apothecia black to dark brown, 0.7–2 mm wide, with prominent margins, sometimes coated with white pruina. Spores with round cells (*Pachysporaria*-type) when mature, 19–31 × 9–16 µm. Spot tests: K+ yellow, KC–, C–, P+ pale yellow. Fairly common, usually on soil, moss, or rock, occasionally on bark, statewide, mostly in coastal areas. Photo from the Channel Is.

Rinodina californiensis

California pepper-spore lichen
Thallus thin, light to dark gray, continuous or rimose. Apothecia black, sessile, lecanorine, 0.3–0.7 mm wide, with whitish margins. Spores almost

R. archaea

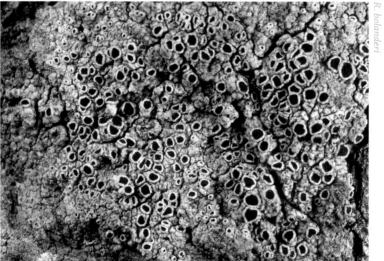
R. bolanderi

Physcia-type, 16–20 × 9–12 μm. Spot tests: K+ yellow, KC–, C–, P– or faint yellow. Occasional on deciduous tree bark, especially twigs, or on wood, statewide, mostly in the Coast Range, rare in the Sierra Nevada. A very similar species, *R. santae-monicae* (not pictured), with a gray to greenish brown thallus, mostly flat apothecia, and different chemistry is also found in the Coast Range of central and southern CA. Also can resemble *R. marysvillensis* (not

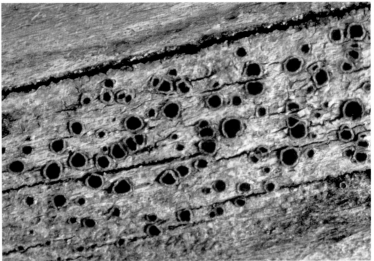

pictured), but that species has apothecia that are usually pruinose. Photo from Six Rivers National Forest.

Rinodina herrei

Herre's pepper-spore lichen

Thalli when young are continuous, becoming rimose and finally areolate, brown to somewhat greenish. Apothecia dark brown to black, pruinose, with distinct margins. Spores have angular cells (*Physcia*-type) at first, later becoming rounded (resembling *Pachysporaria*-type), 16–23 × 8–13 μm. Spot tests negative. On bark, especially of oaks, occasionally on soil or wood, mostly in coastal areas, San Francisco Bay to Mexico. The photo, from the Channel Is., shows a specimen in which the pruina is not obvious.

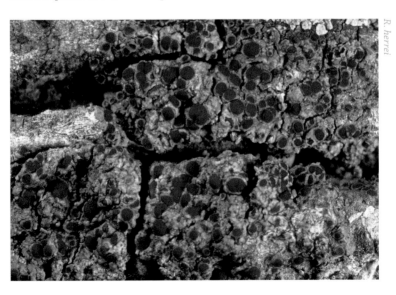

Also of Note

Not described above but relatively common are:

Rinodina badiexcipula, with a thallus of gray to dark brown, discrete areoles; apothecia large (0.5–1 mm), black, with rather thick rims that sometimes have a white or orange pruina and a pigmented proper exciple; spores *Physcia*-type; spot tests: K– or K+ sordid yellow, C–, KC–, P–; on bark of trees and shrubs, statewide except Mojave Desert.

Rinodina bischoffii, with a more or less endolithic thallus (dark gray when visible); apothecia black, 0.35–0.5 mm wide, having a pigmented band around the septum and unthickened walls at the tips; spores broad, *Bischoffii*-type; spot tests negative; on limestone and calcareous sandstone, central and southern coast and mountains.

Rinodina capensis, with a thin thallus of pale gray, slightly convex areoles; apothecia black with thallus-colored margins; spores ellipsoid, *Physcia*-type; spot tests: K+ yellow, C–, KC–, P– or P+ faint yellow; fairly common, and a pioneer species, on twigs and branches of deciduous and coniferous trees, statewide except Mojave Desert.

Rinodina confragosa, which closely resembles *R. capensis,* above, but grows on rock in both coastal areas and mountains.

Rinodina endospora, with a thallus of gray to dark brown, discrete verrucae or irregular areoles; apothecia black, 0.4–0.7 mm wide, with thallus-colored rims and an unpigmented exciple; spores *Dirinaria*-type; spot tests negative; on bark of deciduous trees and shrubs, statewide except Mojave Desert.

Rinodina gennarii, with a thin, rimose to areolate, dark gray to brown thallus; apothecia black, 0.25–0.6 mm wide, with margins the same color as the thallus; spores *Dirinaria*-type; spot tests negative; rather common on maritime rocks statewide, especially those used as bird perches.

Rinodina hallii, with a thin thallus, usually continuous becoming rimose, gray or ocher to pale brown; apothecia large (typically 0.7–0.9 mm wide), biatorine, dark brown to black, and often pruinose; spores *Physcia*-type; spot tests negative; on bark in coastal habitats statewide.

Rinodina intermedia, with a thallus that is usually continuous, light gray-green to brownish; apothecia dark brown to black; common on soil or on moss over soil, often in shady cracks between rocks.

Rinodina laevigata, with a thin, evanescent or sometimes areolate, gray to brown thallus; apothecia black, 0.4–0.9 mm wide, with thallus-colored margins; spores *Physcia*-type; spot tests negative; common on small twigs of deciduous trees, statewide except Mojave Desert.

Sarcogyne

Grain-spored lichens

Crustose lichens with an endolithic, usually invisible, thallus; all that shows are lecideine apothecia on rock. Spores tiny, colorless, 100+ per ascus. Spot tests negative.

Sarcogyne hypophaea
Black grain-spored lichen

Apothecia black, lecideine, 0.3–0.7 mm wide, epruinose, with smooth disks on acid rock; no visible thallus. Inland sites, central to southern CA, most often in southern mountains. Compare with *Lecidea laboriosa* and *Polysporina simplex*. The name includes some lichens previously called *S. privigna*. A similar species, *S. clavus* (not pictured), has apothecia 1–3 mm wide with narrow attachment points. Photo from Joshua Tree National Park.

Sarcogyne regularis
Frosted grain-spored lichen

Apothecia black, or red-brown when wet, but often covered with white to bluish pruina. On calcareous rock, mostly in dry areas but sometimes coastal, central and southern coast and mountains. Photo from a herbarium specimen collected in ON.

Sigridea californica

Gray frost rim lichen

Thallus crustose, white to pale gray. Apothecia gray, lecanorine, heavily pruinose with white rims. Spores fusiform, colorless, 4-celled, 13–15 × 3–4 µm,
8 per ascus. Spot tests: thallus K– or yellowish, CK–, C–, P+ yellow. On bark,
sometimes on wood, rarely on rock, always coastal, statewide, but most common in southern CA. One of the few lichens that is often found on *Eucalyptus*
trees. *Sigridea* and *Dirina* are closely related genera in the same family, common along the California coast. They both have colorless, 4-celled spores and
a very dark hypothecium. They differ significantly in chemistry. *Sigridia* can
also be mistaken for a species of *Lecanora*, but the spores, hypothecium, and
ascus structure are different. Photo from the Channel Is.

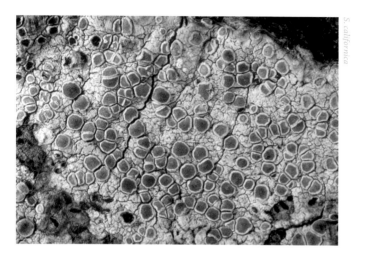

S. californica

Sporastatia testudinea

Copper patch lichen

Thallus crustose, brown, areolate, but usually with somewhat radiating lobes
at the margin and a conspicuous black prothallus. Apothecia black with thin

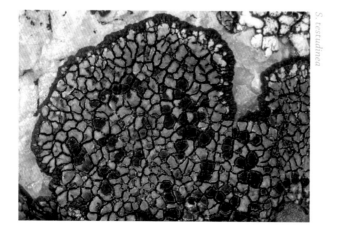

S. testudinea

black rims found between the areoles. Spores ellipsoid to almost round, colorless, 1-celled, several hundred per ascus. Spot tests: cortex and medulla K–, KC+ pink, C+ pink, P–. Resembles a smaller version of *Lecidea atrobrunnea*, although the copper-brown color and general aspect are fairly distinctive. Frequent on rock at higher elevations in inland mountains such as the Sierra Nevada and southern CA coastal ranges. Photo from the Rocky Mountains, CO.

Staurothele
Rock pimples
Crustose lichens with brown to gray thalli and black perithecia. Spores muriform, colorless or brown, 2 per ascus for the species below. Spot tests negative. On rock. Compare with *Verrucaria*, another perithecial genus, and with *Hydropunctaria maura*.

Staurothele areolata
Dry rock pimple
Thallus areolate, dark brown, with convex areoles sometimes becoming almost subsquamulose. Perithecia show in the larger areoles. Rather similar to *S. drummondii* but lacks a prothallus or lobes on the thallus margins and grows in dry locations, central and southern mountains. Photo from near Libby, MT.

S. areolata

Staurothele drummondii
Drummond's rock pimple
Thallus areolate, dark brown or grayish, usually continuous and sometimes becoming almost lobate at the margins, often with a black prothallus. Areoles usually have black perithecia with a well-developed involucrellum around the ostioles. On rocks, especially along streams where they are occasionally under water or on rock faces with frequent seepage, central and southern CA and Sierra Nevada, coast and inland. Photo from the Yukon.

S. drummondii

Staurothele fissa
Lakezone lichen

Thallus dark brown to almost black, continuous or slightly cracked, with black perithecia appearing as bumps in the surface. Occasional, semiaquatic, on rock along lakeshores, sometimes forming a distinct black zone above the water, in inland mountains statewide. Photo from a herbarium specimen collected in QC.

S. fissa

Staurothele monicae
Southern rock pimple

Thallus greenish, gray, or brown, areolate, with areoles packed tightly together. Perithecia black, with the base partly immersed and a conspicuous involucrellum that rises above the areoles. Occasional, on calcareous rock, usually in mountains, central to southern Coast Range and inland mountains. Photo from San Juan National Forest, CO.

Tephromela atra
Black-eye lichen

Thallus crustose, creamy white, and often shiny, thick and areolate when on rock but thinner and more verruculose when on wood or bark. Apothecia

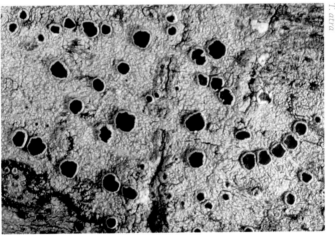

conspicuous, shiny, black, with lecanorine margins. Hymenium purple. Hypothecium yellow. Spores broadly ellipsoid, simple, colorless, 8 per ascus. Spot tests: cortex K+ yellow, KC+ yellow, C–, UV–; medulla K–, KC+ pink-violet, C–, P–, UC+ white. Resembles several species of *Lecanora,* especially *L. cenisia* and *L. gangleoides,* but the color of the hypothecium and hymenium in *T. atra* will distinguish them. Moderately common, usually on rock, but also on wood, or less commonly on bark, statewide, usually toward the coast. Photo from Humboldt Co.

Texosporium sancti-jacobi
Rabbit pellet lichen

A minute crustose lichen with a creamy white granular or verrucose thallus and lecanorine apothecia, 0.5–0.8 mm wide, that have a distinctly greenish, somewhat pruinose mazaedium in the center. Spores brown, with a pseudo-parenchymatous coat of hyphae, 8 per ascus. Spot tests negative. Occasional, on soil and animal pellets, especially those of rabbits, in arid, intermountain

T. sancti-jacobi

sites, central to southern CA. Somewhat resembles species of *Cyphelium* or *Thelomma,* but the spores are distinctive. Photo from a herbarium specimen collected in ID.

Thelomma
Nipple lichens

Crustose lichens with thick thalli, areolate or lobed, gray or yellowish tan. Ascomata are mazaedia, black masses of ascospores embedded in conspicuous spherical warts. Spores ellipsoid to spherical, brown to black, 1- or 2-celled. Mostly on rock, sometimes on hard wood, and mostly coastal. The conspicuously rounded warts (verrucae) with black centers make the genus distinctive.

Thelomma californicum
Lobed nipple lichen

Thallus pale gray, thick and verrucose-areolate in the center with radiating lobes that are pruinose to granular-scabrose at the margins. Mazaedia black,

T. californicum

embedded in cone-shaped verrucae, often with white collars. Spores 2-celled, 15–20 × 10–12 µm. Spot tests: cortex K–, KC+ rose-red, C–, P–; medulla no reactions but UV+ blue-white. The only species of this genus with a lobed thallus. Often on wood, sometimes on rock, in coastal areas statewide but rare or absent south of the Channel Is. Photo from an old fencepost in Pt. Reyes National Seashore.

Thelomma mammosum
Rock nipple lichen

Thallus thick, creamy yellowish to tan or slightly pink, areolate to verrucose, with rounded fertile verrucae 0.8–1.6 mm wide, bearing black mazaedia. Spores spherical, 1-celled, 13–17 µm wide. Spot tests: cortex K–, KC+ pink, C–, P–, UV+ white; mazaedium K+ red, P+ yellow. Fairly common on rocks along the coast, statewide, more common in southern CA. Compare with *T. santessonii*. Photo from Baja CA.

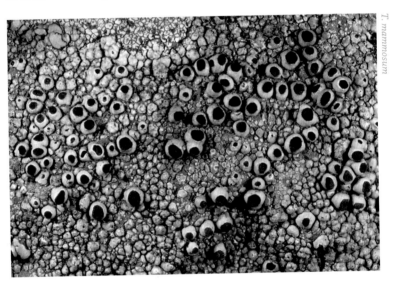

Thelomma occidentale
Western nipple lichen

Thallus lumpy, areolate, pale gray, effuse, without marginal lobes. Verrucae rounded, with black mazaedia that sometimes have a greenish cast. Spores 2-celled, 22–28 × 12–15 µm. Spot tests negative. Rather rare, on wood, especially fences, central and northern CA coast. Larger spores and without the marginal lobes of *T. californicum*. A similar species, *T. ocellatum* (not pictured), has crowded verrucae and often produces dark clusters of isidia, especially when apothecia are absent; spot tests: medulla reacts IKI+ blue; on nutrient-rich wood, such as the tops of fenceposts, statewide, mostly in interior mountain areas. Photo from Sonoma Co.

T. occidentale

Thelomma santessonii

Tan nipple lichen

Similar in appearance to *T. mammosum,* but less common, with a distinctive yellowish tan thallus and larger fertile verrucae, up to 2.5 mm across. Spot tests: cortex KC–; mazaedium K+ red, P+ yellow; UV+ blue-white. Spores spherical, 1-celled, 14–16 μm wide. On coastal rock, central to southern CA. Photo from the Channel Is.

T. santessonii

Thelotrema lepadinum
Bark barnacles

Crustose, with the thallus mostly within the bark substrate, leaving a brownish stain. Apothecia like tiny barnacles or volcanoes, 0.7–1.6 mm wide, with a secondary thin wall sometimes visible within the "volcano." Spores very large, colorless, muriform, mostly 4 per ascus. Spot tests negative. Rather rare, on trees in humid environments, mostly north coast, Mendocino Co. to OR. Photo from Prince William Sound, AK.

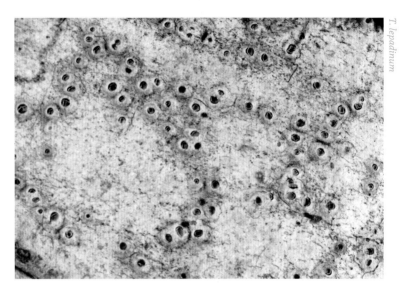

Toninia
Blister lichens

Thalli are usually thick, squamulose or crustose, with flat to strongly convex lobes and verrucae; color is gray to brown and often pruinose; the medulla is white. Apothecia are lecideine. Spores ellipsoid to needle-shaped, colorless, 1–8-celled, 8 per ascus. Spot tests negative for the species below, except as noted. Not common, on soil and rock, often starting as parasitic on other lichens, then becoming free-living.

Toninia ruginosa
Wrinkled blister lichen

Thallus squamulose, with squamules up to 5 mm wide, sometimes forming a continuous surface, olive to reddish brown. Apothecia black but slightly pruinose. Spot tests: epithecium K+ red-brown. Spores 2–8-celled. On soil and rock, often parasitic on cyanolichens. Two subspecies have been recognized based on spore and squamule size, as well as habitat. Photo from a herbarium specimen collected in CO.

Toninia sedifolia

Earth wrinkles, Blue blister lichen

Thallus a heap of irregular squamules and verrucae (warts), dark olive or brown but mostly appearing blue-gray to white because of the pruinose surface. Black, sometimes pruinose apothecia have margins when young but then lose them. Spores cigar-shaped, 2-celled. Usually on calcareous soil in arid areas, often forming rounded lumps, occasionally on rock, scattered statewide, but fairly common in southern mountains and deserts, also in coastal areas. A much rarer species is *T. massata* (not pictured), with flatter squamules, without white margins; in southern mountain areas. Photo from Kootenai National Forest, MT.

Also of Note

A few other species, not pictured here, occur in CA. Two of the more common ones are:

Toninia aromatica, with a gray to yellowish brown squamulose thallus; apothecia reddish; spores 2–4-celled; on soil and rock along the southern coast and in mountains.

Toninia submexicana, with a squamulose thallus that is shiny, dark olive-brown; apothecia grayish; spores 4-celled; on soil and rock in southern deserts, less often near the coast.

Trapelia glebulosa

Pebble lichen

Thallus thin, crustose, dispersed areolate to areolate, pale gray. Apothecia small, rounded, but irregular, brownish, with margins that are somewhat lecanorine but ragged. Spores ellipsoid, colorless, 1-celled, 8 per ascus. Spot tests: K–, KC+ red, C+ pink, P–. Generally rare, on rock, especially pebbles, sometimes on soil, most often on the southern coast. A similar species is *T. coarctata* (not pictured), with a pale, smooth thallus; mostly coastal on rock. The species formerly called *T. involuta* is now included in *T. glebulosa.* Photo from near Lake Placid, NY.

T. glebulosa

Trapeliopsis

Mottled-disk lichens

Crustose lichens with thick thalli that may be granular, areolate, verrucose, or squamulose. Surface is various shades of gray, usually sorediate, sometimes with biatorine apothecia. Spores ellipsoid, colorless, 1-celled, 8 per ascus. Spot tests for the species below: K–, KC+ red, C+ pink, P–.

Trapeliopsis flexuosa
Board lichen

Thallus of dark gray to gray-green areoles, or just consisting of granules that break open when mature, revealing dark greenish soredia. Apothecia occasional, gray to black with thin, distinct margins. Usually on hard wood, such as fence boards, in exposed sites, fairly common statewide, mostly in coastal areas but occasionally in the Sierra Nevada and inland mountains. Compare with *T. granulosa,* which has a thallus and soralia of lighter color; typically on softer dead wood. Photo from Mendocino Co.

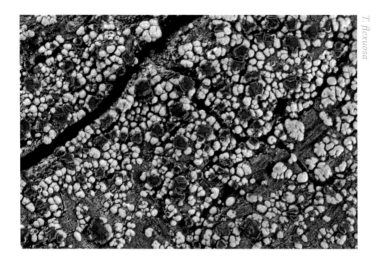

T. flexuosa

Trapeliopsis glaucopholis
Scaly mottled-disk lichen

Thallus squamulose, pale gray, with lobes 0.3–1 mm wide, often irregular in size and shape, appressed to ascending. Apothecia flat or convex, reddish brown to dark gray or almost black, with or without margins. Usually with

T. glaucopholis

immersed, brown pycnidia. On soil, sometimes on rock, statewide, coast and inland mountains. This is what most North American lichenologists previously called *T. walrothii,* a European species. Photo from Memaloose State Park, OR.

Trapeliopsis granulosa
Mottled-disk lichen

Thallus pale gray, sometimes greenish or pinkish, consisting of cracked areoles or, more frequently, granular to hemispherical verrucae that often become sorediate, with coarse, greenish to white soredia. Apothecia occasional, pink to tan or even black, often mottled, 0.4–1.5 mm wide, with or without margins. Fairly common, on acidic soil or wood, especially burned wood, rarely on rock, statewide. Compare with the darker *T. flexuosa.* Photo from Mt. Hood National Forest, OR.

T. granulosa

Vahliella
Brown shingle lichens, Mouse lichens

Species in this genus closely resemble those in *Fuscopannaria* and also contain cyanobacteria. See description under that name.

Vahliella leucophaea
Thallus squamulose with dark, gray-brown, overlapping, and fairly thick squamules; sometimes with a black prothallus. Apothecia dark brown to almost black, with or without lecanorine margins or with partial ones; without soredia or isidia. Spores ellipsoid, colorless, smooth, 1-celled, 8 per ascus. Spot tests negative. Occasional, on shady rock, especially where damp, mostly in the Coast Range and northern mountains. Compare with *Fuscopannaria* and *Pannaria.* Another species, *V. californica* (not pictured), has very thick, brown squamules and brown apothecia; on rock in the Sierra Nevada and occasionally on the coast. Photo from San Juan National Forest, CO.

Verrucaria

Speck lichens

Crustose lichens whose ascomata are perithecia. Thalli of the most common species are dark gray to black, but others can be white, gray, or brown, and some are endolithic. Spores ellipsoid to spherical, without a gelatinous halo, colorless, smooth, 1-celled, 8 per ascus; no paraphyses seen in the perithecia. Spot tests negative. Some species can be found in maritime tidal zones or in mountains on stones washed by freshwater or on dry rocks. Compare with *Staurothele* and *Hydropunctaria*. Related genera (also lacking paraphyses), not included here, are *Polyblastia* and *Thelidium;* they differ in spore septation. The genus is difficult to identify, even for experienced lichenologists, and a number of species in addition to those described below occur in CA.

Verrucaria nigrescens

Smooth speck lichen

Thallus crustose, cracked-areolate, rather smooth, dark olive to brownish black, often with a black prothallus, epruinose. Perithecia entirely embedded

in the thallus or slightly protruding, appearing as slight bumps. Occasional on rock, most often limestone, or on concrete, rarely on siliceous rock, in mostly coastal habitats, probably statewide. Photo from southern ON.

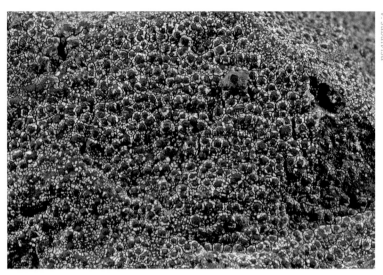

V. subdivisa

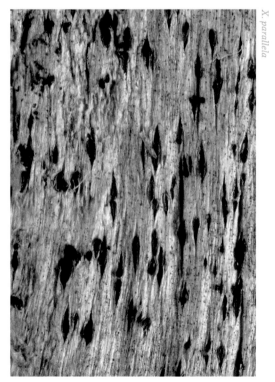

X. parallela

Verrucaria subdivisa
Cracked speck lichen
Thallus strongly areolate, surface gray to brown with gray to white pruina, with a thick carbonaceous basal layer, thinning at the margins, and usually with a black prothallus. Perithecia black, immersed, 1 or 2 per areole. On rock, central and southern Coast Range. Photo from Point Lobos, Monterey Co.

Xylographa parallela
Black woodscript
Thallus crustose, entirely within the wood substrate, with only a gray stain showing on the surface and elongated black lirellae that follow the wood grain. Spores often few or immature, ellipsoid, simple, 8 per ascus. Spot tests: most parts negative but medulla under the lirellae usually K+ yellow, KC–, C–, P+ yellow-orange, IKI+ blue. Occasional on rotting wood such as logs and stumps, statewide, mostly in humid, shady montane sites. Another, rather rare species, *X. vitiligo* (not pictured), has brownish soralia, brown lirellae, and the same spot tests; north coast and Sierra Nevada. Also compare with species of *Opegrapha*. Photo from a herbarium specimen collected in MB.

Mushroom Lichens

The two species listed below are lichenlike associations formed with basidiomycete fungi, the kinds that normally form mushrooms and rarely form lichens. Almost all lichens have an ascomycete fungal partner.

Lichenomphalia umbellifera
Greenpea mushroom lichen

Thallus is a mass of tiny green globules, out of which arise small, orange-tan, gilled mushrooms that lack veils or basal cups. Unlike most lichens, *Lichenomphallia* is a basidiomycete lichen. Each green globule is a colorless,

L. umbellifera

spherical, fungal envelope containing a few algal cells (the green alga *Coccomyxa*). Infrequently found, perhaps due as much to being unrecognized as a lichen as to its actual occurrence. Unlike most other lichens, the fruiting body is seasonal, like other mushrooms. On damp, sloping soil such as on road-cuts, or on rotting wood. Range largely unknown, but probably more coastal and more northern. Compare with *Multiclavula,* another basidiomycete lichenlike association. Photo from coastal OR.

Multiclavula corynoides
Club-mushroom "lichen"

Without a true thallus, simply a layer of green algae growing over a substrate, usually decaying wood or soil, with erect, pale yellowish, club-shaped, basidiocarp fruiting bodies. Since it doesn't form an actual thallus, with fungal hyphae and associated algae forming a new entity unlike either partner, it cannot be considered a fully developed lichen, but it does have a lichenlike symbiotic character. No lichen substances. Compare with *Lichenomphalia,* which forms gilled mushrooms and is more lichenized. On mossy soil, mostly Coast Range. The photo, from Monterey Co., shows *Peltigera venosa* at the left, which I have observed more than once growing alongside *Multiclavula.*

M. corynoides

Appendix 1: Spot Tests

Many chemicals are found in lichens, and their presence or absence is often important for identification. For this reason the species descriptions often include information about spot tests, listing the colors that lichen tissues turn when various chemical reagents are applied.

The basic techniques are simple: one uses a fine pipette (like a capillary tube used for blood tests, but drawn to a fine point) to apply a tiny amount of reagent to the lichen, often to a specific part of it, such as the medulla, which one can expose with a razor blade. Since you usually want to test only a minute portion of the lichen, and to use as little reagent as possible, the tests are often best done under a dissecting microscope so that you can distinguish among reactions in the cortex, the medulla, and the soralia.

The common reagents are described below:

K: a 10% solution of KOH, potassium hydroxide. It will remain effective for six months to a year.

A K test on the white medulla of *Parmotrema crinitum,* showing the yellow reaction.

C: a strong solution of NaClO, sodium hypochlorite, commonly known as bleach. Undiluted laundry bleach will do for this, but avoid ones with additives such as scents. You can store a small quantity in a dropper bottle, but replace it from the large bottle of bleach after a week or two.

P: a solution of *para*-phenylenediamine. One makes the solution by adding a drop of 70% ethyl or isopropyl alcohol to a few crystals of PD on a microscope slide or in a vial. PD must be obtained from a chemical supply house. The solution keeps its effectiveness for only about an hour, and it is sensitive to light. A longer-lasting form of P is the mixture called Steiner's solution: dissolve 10 g sodium sulfite in 90 ml of water, add 5–6 drops of liquid detergent, then 1 g of PD. Stir for a few minutes, then filter out the undissolved crystals; the solution, which should be pink, should last for some months and can be quite useful even if the reactions it produces are a bit weaker than those with the alcohol solution. It gradually turns darker and darker purple.

Less commonly used reagents are:

IKI: a solution of iodine, made by mixing 1.5 ml of Lugol's solution (obtainable from a supply house or online) with 18.5 ml of pH 11.0 buffer. Or one can mix a solution from scratch by first preparing potassium iodide solution (4 gm of KI dissolved in 96 ml of distilled water) and then adding 1.5 gm of iodine crystals (extremely toxic!). The mixture fades in light and should be stored in an amber bottle. After a month or so, small quantities are often too pale to be useful. Use 1/3 strength for ascus tips.

K/I: refers to staining with iodine after KOH, used in microscopic study of ascus tips.

N: a 35% solution of nitric acid ("concentrated" acid is normally supplied as a 70% solution, so diluting that by half yields a 35% solution), useful in telling species of *Melanelia* from brown species of *Xanthoparmelia* (i.e., "*Neofuscelia*") and testing color changes in certain lichen pigments.

In most tests the reagents are applied singly, but in the case of a "KC" test, one first applies a drop of K, then a drop of C while the lichen is still damp from the K.

For dark-pigmented lichens, especially those in the genus *Bryoria*, the "filter paper" method can be very useful. Put a few strands of the lichen on a 1 cm² square of filter paper (over a glass plate or microscope slide). Then flood the lichen with a drop or two of reagent, and the reactive lichen substances will flow out onto the filter paper, where their colors can be seen, or, under a dissecting microscope, they can be seen within the lichen strands, which will be rendered semitransparent by the Steiner's solution.

All of these chemicals need care in handling; PD, especially, is very poisonous; nitric acid and iodine crystals are highly reactive and dangerous to exposed skin, but only tiny amounts of them are needed. All will stain or otherwise damage skin, clothing, or surfaces.

The "UV" test involves looking at the lichen under ultraviolet light rather than using chemicals. Ultraviolet light sources can be short-wave or long-wave, but only long-wave is useful in testing lichens directly. More complete information about these methods can be found in *Lichens of North America* and other sources (see especially A. Orange, P. W. James, and F. J. White, "Microchemical Methods for the Identification of Lichens," *British Lichen Society* [2001]), and good guidance can be found by participating in workshops of the California Lichen Society.

Appendix 2: Recent Name Changes

Lichen taxonomy is in a state of rapid change, mostly due to molecular studies and cladistics, and many names have changed. Most of the changes (for species mentioned in this book) since the publication of *Lichens of North America* are listed below.

Some of the "new" names below are old ones but now include species formerly called something else, for example, *Bryoria fremontii.* Frequently there is not a one-to-one correspondence between an older name and a new one; for example, the lichen formerly called *Pyrrhospora cinnabarina* was split into several species, with *Ramboldia cinnabarina,* used for sorediate specimens, and nonsorediate ones (pictured in this book) becoming *Ramboldia gowardiana.* Likewise, *Punctelia subrudecta* was split into three species; the common one in CA is *P. jeckeri.*

The new names are not necessarily synonyms of the old ones; some of the old names are still valid but have been misapplied in some way, perhaps only in western North America, and should not be used for lichens found in CA.

Old	*New*
Aspicilia desertorum	*Circinaria arida*
Aspicilia contorta	*Circinaria contorta*
Bryoria tortuosa	*Bryoria fremontii*
Buellia subalbula	*Buellia maritima*
Candelariella terrigena	*Candelariella citrina*
Cladonia cervicornis subsp. *verticillata*	*Cladonia concinna*
Dimerella lutea	*Coenogonium luteum*
Fuscopannaria leucophaea	*Vahliella leucophaea*
Hafelia disciformis	*Buellia disciformis*
Hubbsia parishii	*Schizopelte parishii*
Lecidella eleochromoides	*Lecidella asema*
Lepraria lobificans	*Lepraria finkii*
Leprocaulon microscopicum	*Leprocaulon americanum*
Leptogium corniculatum	*Leptogium palmatum*
Melanelia (several species)	*Melanelixia, Melanohalea*
Melanelixia glabra	*Melanelixia californica*

Old	New
Omphalina umbellifera	*Lichenomphalia umbellifera*
Opegrapha atra	*Arthonia atra*
Opegrapha brattiae	*Lecanographa brattiae*
Parmotrema chinense	*Parmotrema perlatum*
Phaeophyscia cernohorskyi	*Phaeophyscia hirsuta*
Placidium lacinulatum	*Clavascidium lacinulatum*
Placidium tuckermanii	*Placidium arboreum*
Punctelia subrudecta	*Punctelia jeckeri*
Pyrrhospora cinnabarina	*Ramboldia gowardiana*
Sphaerophorus globosus	*Sphaerophorus venerabilis*
Sphaerophorus globosus var. *gracilis*	*Sphaerophorus tuckermanii*
Teloschistes californicus	*Seirophora californica*
Trapelia involuta	*Trapelia glebulosa*
Trapeliopsis walrothii	*Trapeliopsis glaucopholis*
Usnea arizonica	*Usnea intermedia*
Usnea wirthii	*Usnea flavocardia*
Verrucaria compacta	*Heteroplacidium compactum*
Verrucaria maura	*Hydropunctaria maura*
Xanthoria (several species)	*Xanthomendoza*

Genus names incorporated into an existing genus

Old	Now in
Cavernularia (both species)	*Hypogymnia*
Cladina (all species)	*Cladonia*
Neofuscelia (all species)	*Xanthoparmelia*
Roccellina (all N. American species)	*Dendrographa*

Glossary

– – – – – – –

Adnate
Attached tightly to the surface.

Algae (sing. alga)
Photosynthetic organisms with chloroplasts and nuclei, usually green in color.

Algal layer
A layer of algal cells (or cyanobacteria) just below the upper cortex of layered (stratified) lichens.

Amphithecium
The part of a lecanorine apothecium outside the exciple, constituting the thalline margin, typically containing thallus-like tissues such as algae, a cortex, and sometimes a medulla.

Apothecium (pl. apothecia)
A disk or cup-shaped ascoma, or fruiting body, usually with an exposed hymenium. In a vertical section of an apothecium, the more or less clear, palisade-like layer close to the disk surface is the hymenium. It contains the asci and the paraphyses. The uppermost portion of the hymenium, formed by the tips of paraphyses, is the epihymenium (i.e., "on top of" the hymenium). Below the hymenium is a thin or thick, often pigmented tissue called the hypothecium (*hypo* meaning "below" and *thecium* being another term for hymenium). Between the hypothecium and hymenium is often a thin, poorly defined (rarely thick) tissue called the subhymenium (with much the same meaning as hypothecium, but situated above it). The exciple is tissue below and to the sides of the hymenium and hypothecium; it forms the margin of biatorine and lecideine apothecia. The exciple is much reduced in lecanorine apothecia, where an additional marginal tissue is formed external to the exciple that resembles the thallus in structure: the amphithecium.

Appressed
Closely adnate, flattened.

Areolate
A lichen thallus that is broken into patches (areoles), often like tiny tiles.

Ascoma (pl. ascomata)
The fruiting body of an Ascomycete fungus, inside of which are asci that contain the ascospores.

Ascus (pl. asci)
The saclike structure in an Ascomycete fruiting body that produces ascospores.

Axil
The angle between two branches.

Biatorine
A type of apothecium with a pale or clear (uncarbonized) margin without photobiont cells, usually resembling the disk in color.

Bischoffii-type
A spore type characterized by dark brown, 2-celled spores having a pigmented band around the septum and unthickened walls except at the septum.

Blastidium (pl. blastidia)
A granule-sized thallus fragment that buds off from the margin.

Bullate
A thallus with inflated, swollen areoles.

Calcareous
Rock or soil that is high in calcium carbonate, like limestone; basic rather than acidic.

Capitulum (pl. capitula)
The round or cup-shaped apothecium on the tip of a stalk in the stubble lichens, e.g., *Calicium* and *Chaenotheca*.

Cephalodium (pl. cephalodia)
A small growth like a gall, containing cyanobacteria, that occurs in some lichens with green algae as the main photobiont.

Chemotype
A form of a particular species that has distinctive chemistry but is otherwise the same.

Cilia (sing. cilium)
Small, hairlike appendages on the margins of the thallus or apothecia of many lichens, e.g., *Parmotrema* and *Physcia tenella*.

Clavate
Club-shaped.

Conidia (sing. conidium)
The asexual spores found in pycnidia.

Consoredia
A spherical conglomeration of soredia that looks like a large granular soredium.

Coralloid
With minutely branched cylindrical outgrowths, like coral.

Cortex
The "skin" of a stratified lichen; i.e., the outer layer of fungal tissue that protects the interior tissues.

Corticolous
Growing on bark.

Crustose
A lichen growth form in which the thallus lies flat on the substrate, in contact with it at all points, is without a lower cortex, and is not removable without also removing some substrate.

Cyanobacteria (sing. cyanobacterium)
Photosynthetic organisms closely related to bacteria, formerly called blue-green algae, that form the photobiont in some genera of lichens or are found in the cephalodia of others. Cyanobacteria "fix" atmospheric nitrogen and thus are important in an ecosystem's nitrogen cycle.

Cyphellae (sing. cyphella)
Round pores in the lower surface of some lichens, notably in *Sticta,* lined with loosely packed, spherical cells.

Dichotomous
Branching into two lobes or branches like the letter "Y"; the forking branches may be equally or unequally thick, but the shape is approximately symmetrical.

Dirinaria-type
A spore type characterized by dark brown, 2-celled spores with angular locules as in the *Physcia*-type, but with the central septum area swelling with the application of KOH.

Effigurate
With a well-defined, usually lobed, margin.

Endemic
Found only in a certain area, e.g., California.

Epihymenium
Uppermost portion of the hymenium, formed by the tips of paraphyses. The epihymenium is the visible, central portion of an apothecium when seen from above. (See also the description under Apothecium.)

Epruinose
Without pruina.

Exciple
Part of the apothecium outside and below the hypothecium, forming the apothecial margin in lecideine and biatorine apothecia, reduced in lecanorine apothecia. In reference to perithecia, it is sometimes termed an excipulum. (See also the description under Apothecium.)

Farinose
Like fine powder; referring to soredia.

Fibril
A short side branch in the genus *Usnea*.

Foliose
Leaflike, referring to the lichen growth form with prominent lobes and distinct upper and lower surfaces; basically, two-dimensional.

Fruticose
A lichen growth form that can be shrubby or pendant or stalked, without clear upper or lower surfaces; basically three-dimensional.

Fungus (pl. fungi)
A form of life that is not photosynthetic and thus must get its energy by parasitizing live organisms or digesting dead ones or (as in the case of lichens) by forming a symbiosis with an alga or cyanobacterium. Fungi form tissues made of hyphae and reproduce by means of spores. They belong to the Kingdom Fungi; i.e., they are neither plants nor animals.

Fusiform
Spindle- or cigar-shaped; elongated and narrowing at each end.

Holotype
The specimen upon which a new species is based when it is originally described.

Hymenium
The portion of an ascoma that produces the spores, consisting of a layer of asci and paraphyses and related structures. (See also the description under Apothecium.)

Hyphae (sing. hypha)
Fungal filaments, usually elongated but sometimes divided into short, angular or almost round cells, that make up the mass of fungal tissue.

Hypothallus
A layer of fungal tissue that underlies the main thallus, sometimes extending out at the edge and visible as a border of a different color; often black, but may be white or other colors. Sometimes used as synonymous with "prothallus," but the latter is assumed to represent an earlier developmental stage and is generally considered to be associated with crustose lichens, with "hypothallus" being used for foliose or squamulose species, e.g., *Pannaria rubiginosa*.

Hypothecium
The layer of tissue between the hymenium (and subhymenium) and exciple in an ascoma, often distinctly textured or colored but otherwise similar to the exciple. (See also the description under Apothecium.)

Imbricate
Overlapping like shingles; usually used to describe the thalli of some squamulose lichens.

Involucrellum
The carbonaceous cap or covering outside the excipulum (see Exciple), present on many perithecia.

Isidia (sing. isidium)
Minute corticate outgrowths from the thallus containing photobiont cells; they detach easily and thus act as vegetative propagules.

Laminal
On the upper surface of a thallus (as opposed to along the margins).

Lecanorine
A type of apothecium typical of the genus *Lecanora* that has a margin containing photobiont cells and is similar in color and texture to the surrounding thallus.

Lecideine
A type of apothecium typical of the genus *Lecidea* that has a margin without photobiont cells and with the exciple at least partly carbonized, forming a black apothecial margin.

Lichenicolous
Growing on or in a lichen, used in reference to fungi or lichens that are partly or completely parasitic on the host lichen.

Lirella (pl. lirellae)
A type of apothecium that is elongated, sometimes branched; typical of the genera *Graphis* and *Opegrapha*.

Lobe (lobate)
An elongated or rounded extension of a thallus, measured at the widest point.

Lobules
Minute lobes, like tiny scales, growing out of a thallus margin or from the surface; they have tissue resembling the thallus from which they emerge.

Locule
The cell cavity inside an ascospore.

Maculae (maculate)
Blotches or spots on the thallus that are caused by breaks in the photobiont layer beneath the cortex; generally they are pale and irregular in shape, or reticulate.

Margin (marginal)
The edge of a thallus or ascoma.

Mazaedium
A powdery mass of ascospores and paraphyses, usually dark, formed by the breakdown of asci in some lichens, typical of the genus *Calicium* and its relatives.

Medulla
An internal layer of fungal tissue in the thallus, or in a lecanorine apothecium, composed of loose strands of hyphae; most often white, but sometimes yellowish or otherwise pigmented.

Muriform
Spores that are divided into cells both longitudinally and crosswise, resembling bricks in a wall.

Mycobiont
The fungal component of a lichen; as opposed to photobiont.

Ostioles
The round pores in pycnidia and perithecia through which conidia or ascospores, respectively, escape.

Pachysporaria-type
A spore type characterized by dark brown, 2-celled spores with almost circular locules and very thick, evenly thickened walls.

Papillae (papillate)
Tiny rounded cortical bumps without medullary tissue on the surface of some lichens, especially in the genus *Usnea*.

Paraphyses (sing. paraphysis)
Fungal filaments, sometimes branched, that are associated with asci in the hymenium. (See also the description under Apothecium.)

Peltate
Fastened more or less centrally with a single holdfast; the same as "umbilicate" but on a smaller scale.

Perithecia (sing. perithecium)
Fruiting bodies (ascomata) that are flask-shaped, often embedded in thallus tissue and opening by pores at the summit. Typical of some genera such as *Dermatocarpon*.

Photobiont
The photosynthetic component of a lichen; most often a species of algae but, in some lichens, a cyanobacterium.

Physcia-type
A spore type characterized by dark brown, 2-celled spores with angular locules and unevenly thickened walls that are thicker at the tips of the spore than they are at the sides.

Physconia-type
A spore type characterized by brown, 2-celled spores, typically with thin, fairly uniform walls at the spore tips and a thickened septum.

Podetium (pl. podetia)
A vertical stalk, branched or unbranched, typical of the genus *Cladonia*, made up of lower apothecial tissues, sometimes with a cortex and algal layer, but sometimes not, as in *Baeomyces* (a genus that is rare in CA, not included in this book).

Polarilocular
Spores with 2 cell cavities (locules) at the opposite "poles" of a spore, divided by a thick septum with a narrow canal passing between. Found in *Caloplaca* and its relatives.

Prosoplectenchyma
A type of fungal tissue consisting of long, coalesced cells, often with thick walls. Compare with pseudoparenchyma.

Prothallus
A fringe of purely fungal tissue, white, black, or pigmented, usually seen at the edge of certain crustose lichens (e.g., *Placynthium nigrum*), but also visible between the areoles of others, such as *Rhizocarpon*. See also Hypothallus.

Pruina (pruinose)
A crystalline deposit made up of chemical material as well as dead cells, usually white or pale gray, on the surface of many lichens or on their apothecia. It can make many lichens appear paler than they would otherwise look.

Pseudocyphellae (sing. pseudocyphella)
Tiny white dots on the surface of a thallus where a break in the cortex allows some hyphae from the medulla to come through.

Pseudoparenchyma
A type of fungal tissue consisting of short, rounded to squarish cells, in branched, irregular filaments. Also called paraplectenchyma. Compare with prosoplectenchyma.

Pycnidia (sing. pycnidium)
Round or urn-shaped bodies producing conidia, appearing as small black dots on a thallus.

Reticulate
Netlike.

Rhizine
Hairlike extensions of the lower surface of a thallus that hold the lichen to its substrate; rhizines may be simple, branched, or brushlike, short or long, usually the same color as the lower cortex.

Rimose
A surface that is minutely cracked; used generally to describe some crustose lichens.

Rugose
A surface with rounded wrinkles or ridges.

Scabrose
A minutely roughened surface, usually due to accretions of dead cortical cells.

Schizidia
Irregular fragments of the upper layers of a thallus containing cells from the cortex and photobiont layer that can break off and become vegetative propagules.

Septum
A cross-wall in the cell of a spore or hyphal tissue.

Sessile
Sitting on the surface without a stalk, as in some types of apothecia.

Siliceous
A type of rock high in silica and without calcium; siliceous rock such as granite is acidic rather than basic.

Soralium (pl. soralia)
A place in the thallus where soredia (and sometimes isidia) are produced; soralia may be various shapes and occur along the margins or on the surface of the thallus, marginal or laminal, respectively.

Soredium (pl. soredia)
A kind of vegetative propagule consisting of fungal hyphae surrounding a few algal cells; soredia may be powdery, farinose, or granular, produced in soralia or distributed over the thallus surface or along the margins.

Spore
A reproductive body, typical of fungi, usually used here as synonymous with ascospores (produced in asci); may be single or multicelled.

Squamule (squamulose)
A small, scalelike lobe that lifts up at the edges, typical of the thalli of genera such as *Psora,* or of the basal portion of most *Cladonia* species.

Squarrose
Resembling tiny bottlebrushes, with a main stem and perpendicular side branches; usually referring to rhizines.

Stereome
The cartilaginous, cylindrical structure internal to the medulla that provides supporting tissue within the podetia of *Cladonia* species.

Stipe, stipitate
A stalk; often used in reference to hypothecial tissue extending into the thallus, sometimes elongating and lifting the apothecium.

Stratified
Layered, referring to lichen thalli that have an upper cortex, photobiont layer, medulla, and usually a lower cortex.

Subhymenium
A layer just below the hymenium and above the hypothecium. (See also the description under Apothecium.)

Subsp.
Abbreviation of subspecies.

Substrate
The material on which a lichen grows, usually bark, wood, or rock, but sometimes moss or other vegetative debris, bone, other lichens, or artificial materials such as metal, glass, plastic, or fabric; in rare instances on living animals.

Thallus (pl. thalli)
The main vegetative body of a lichen, containing both mycobiont and photobiont.

Tholus
The thickened tip of an ascus.

Tomentum (tomentose)
A kind of fine hair, like fuzz, on the surface of some lichens, made up of colorless hyphae.

Tubercule (tuberculate)
A wartlike outgrowth of some lichens, containing some medullary tissue.

Type specimen
A specimen used to define a species; there are a number of different sorts of type specimens that taxonomists distinguish.

Umbilicate
Attached to the substrate by a single holdfast (umbilicus), as in the genus *Umbilicaria*.

Variety (abbr. "var.")
A taxonomic category within a species that distinguishes genetic variations considered to be less significant than those between subspecies.

Verrucose
With a warty, rough surface consisting of verrucae (verruca, sing.), which means "warts."

Verruculose
Like verrucose, but with finer bumps.

Wood
Dead wood, without bark; can be hard or partly decayed and soft.

Bibliography and Resources for Further Study

The resources below are listed by author. Some have direct relevance to the study of California lichens and others are more general references. They include only the more recent resources and are far from a complete list.

Books

Brodo, Irwin M., Sylvia Duran Sharnoff, and Stephen Sharnoff. *Lichens of North America.* New Haven: Yale University Press, 2001.

Hale, Mason E. Jr., and Mariette Cole. *Lichens of California.* Berkeley: University of California Press, 1988.

Hinds, James W., and Patricia L. Hinds. *The Macrolichens of New England.* Bronx: The New York Botanical Garden Press, 2007.

McCune, Bruce, and Linda Geiser. *Macrolichens of the Pacific Northwest,* 2nd ed. Eugene: Oregon State University Press, 2009.

McCune, Bruce, and Trevor Goward. *Macrolichens of the Northern Rocky Mountains.* Eureka: Mad River Press, 1995.

Nash, Thomas H. III, ed. *Lichen Biology.* Cambridge: Cambridge University Press, 2008.

Nash, Thomas H. III, Bruce D. Ryan, Corinna Gries, and Frank Bungartz, eds. *Lichen Flora of the Greater Sonoran Desert Region,* 3 vols. Tempe: Arizona State University Press, 2001–2007.

Purvis, William. *Lichens.* Washington, DC: Smithsonian Institution Press in association with the Natural History Museum, London, 2000.

St. Clair, Larry L. *A Color Guidebook to Common Rocky Mountain Lichens.* Provo: M. L. Bean Life Science Museum, Brigham Young University, and the US Forest Service, 1999.

Journals

Some of these are available online.
Bulletin of the California Lichen Society
The Bryologist
Evansia

The Lichenologist
Mycotaxon
Opuscula Philolichenum

Online Resources

Consortium of North American Lichen Herbaria. http://lichenportal.org/portal/.

Esslinger, Theodore L. "A Cumulative Checklist for the Lichen-Forming, Lichenicolous and Allied Fungi of the Continental United States and Canada." http://www.ndsu.edu/pubweb/~esslinge/chcklst/chcklst7.htm.

Goward, Trevor, et al. Ways of Enlichenment. Lichens of Western North America. http://www.waysofenlichenment.net/.

Harris, Richard, and Doughlas Ladd. Lichens of the Ozarks. http://www.nybg.org/bsci/lichens/ozarks/.

May, Philip F., Irwin M. Brodo, and Theodore L. Esslinger. "Identifying North American Lichens: A Guide to the Literature." Farlow Herbarium, Harvard University, rev. 2002. http://www.huh.harvard.edu/collections/lichens/guide/guidetoliterature.html. [General lichen references, works for beginners, lichen keys online, and a lichen bibliography by genus.]

Stephen Sharnoff Photography. Lichens Home Page. http://www.sharnoffphotos.com/lichens/lichens_home_index.html.

Tucker, Shirley C., and Bruce D. Ryan. "Revised Catalogue of Lichens, Lichenicoles, and Allied Fungi in California." *Constancia* 84, 2006, University and Jepson Herbaria, University of California, Berkeley. http://ucjeps.berkeley.edu/constancea/84/.

United States Forest Service. National Lichens & Air Quality Database and Clearinghouse. http://gis.nacse.org/lichenair/.

Index

- - - - -